6th International Conference on Current Issues of Continuum Mechanics and Celestial Mechanics 2016 (CICMCM 2016)

Journal of Physics: Conference Series Volume 919

Tomsk, Russia
16 - 18 November 2016

Editor:

Maxim Orlov

ISBN: 978-1-5108-5362-1
ISSN: 1742-6588

Printed from e-media with permission by:

Curran Associates, Inc.
57 Morehouse Lane
Red Hook, NY 12571

Some format issues inherent in the e-media version may also appear in this print version.

Printed by Curran Associates, Inc. (2018)

For permission requests, please contact the Institute of Physics
at the address below.

Institute of Physics
Dirac House, Temple Back
Bristol BS1 6BE UK

Phone: 44 1 17 929 7481
Fax: 44 1 17 920 0979

techtracking@iop.org

Additional copies of this publication are available from:

Curran Associates, Inc.
57 Morehouse Lane
Red Hook, NY 12571 USA
Phone: 845-758-0400
Fax: 845-758-2633
Email: curran@proceedings.com
Web: www.proceedings.com

Table of contents

Volume 919

6th International Conference Current issues of continuum mechanics and celestial mechanics (6thCICMCM)

16–18 November 2016, Tomsk, Russian Federation

Accepted papers received: 27 October 2017
Published online: 1 December 2017

Invited

Oral

IOP Conf. Series: Journal of Physics: Conf. Series **919** (2017) 011001 doi:10.1088/1742-6596/919/1/011001

Preface

This book contains the papers of the VI International Conference Current issues of continuum mechanics and celestial mechanics – CICMCM'2016 held in Tomsk (Russia), November 16–18, 2016. The first conference was organized in 2010 within the framework of the Second All-Russian Festival of Science at Tomsk State University. More than 750 researchers from scientific organizations of Russia, Algeria, Poland, Belarus, Ukraine, Canada, Germany, Serbia, France and Kazakhstan participated in the conference since its foundation. Almost all National Research Universities of Russia, including Russian Federal Nuclear Centers were represented. Previous papers were published in *Proceedings of Tomsk State University* and *Tomsk State University Journal of Mathematics and Mechanics.*

The conference themes addressed novel and advanced topics on Continuum Mechanics and Celestial Mechanics with focus on gas mechanics, hydrodynamics, aerodynamics, elasticity theory, plasticity theory, deformable solid mechanics, mechanics of heterogeneous and functional materials, movement of bodies in space, liquid and gaseous media, strength, survivability and destruction of materials and structures, technical mechanics of living systems, multiscale structure and texture and ballistics.

The organizers of the conference were Scientific Research Institute of Applied Mathematics and Mechanics and Faculty of Physics and Engineering of Tomsk State University. There are good reasons to believe that the conference fulfilled its goals.

We would like to take this opportunity to thank the authors and all reviewers for their generous efforts.

A sincere thanks to all participants, sponsors and institutions for their great support towards a successful CICMCM'2016.

Special thanks to the Russian Foundation for Basic Research and the local committee of the conference.

We are always glad to see you at Tomsk State University.

Editor-in-Chief
Maxim Orlov

Tomsk, Russia, 2016

IOP Conf. Series: Journal of Physics: Conf. Series **919** (2017) 011002 doi:10.1088/1742-6596/919/1/011002

6th International Conference Current issues of continuum mechanics and celestial mechanics (6^{th}CICMCM) 2016

IOP Conf. Series: Journal of Physics: Conf. Series **919** (2017) 011003 doi:10.1088/1742-6596/919/1/011003

6th International Conference Current issues of continuum mechanics and celestial mechanics (6thCICMCM) 2016

Co-Chairs:
Anatoly Glazunov, Ph.D
Ernst Shrager, Ph.D
Maxim Orlov, Ph.D

Organizer:

National Research Tomsk State University

International Advisory Committee:
Leopold Kruzska, Professor, Military University of Technology (Poland),
Nikita Fomin, Professor, Ph.D., A.V. Luikov Heat and Mass Transfer Institute of NASB (Belarus),
K. Kusainov, Professor, Ph.D., The Karaganda State University of the name of academician E.A. Buketov (Kazakhstan),
Ahmed Brara, Ph.D., CNERB Research Center (Algeria),
Momchilo Milinovich, professor, Ph.D., University of Belgrad (Serbia),
Hussam Ali Khalaf, PhD Marshes Research Center, University of Thi-Qar (Iraq),
Martina Drdlova, PhD, Research Institute for Building Materials (Czech Republic),
Pascal Forquin, Prof. Dr., Universitй Joseph Fourier (France),
Yulie Bai, Ph.D, Beijing University of Technology (China),
Abraham Christian, Ph. D. National University of Singapore (Singapore),
U. Johnson Alengaram, Prof. Dr., University of Malaya (Malasia),
Iqbal M.A. Ph.D. Dr. Indian Institute of Technology Roorkee (India),
Al. Filkov, Ph.D.The University of Melbourne (Australia)

Conference Committee:
Vladimir Arhipov, National Research Tomsk State University (TSU),
Tatiana Bordovitsina, National Research Tomsk State University (TSU),
Vladimir Bimatov, National Research Tomsk State University (TSU),
Vladimir Butov, National Research Tomsk State University (TSU),
Aleksandr Vasilev, Peter the Great St.Petersburg Polytechnic University (SPbPolyTechU),
Victor Glazyrin, National Research Tomsk State University (TSU),
Aleksandr Ishchenko, National Research Tomsk State University (TSU),
Sergey Kul'kov, National Research Tomsk State University (TSU),
Vladimir Skripnyak, National Research Tomsk State University (TSU),
Aleksandr Shvab, National Research Tomsk State University (TSU),
Gennadiy Shrager, National Research Tomsk State University (TSU)

IOP Conf. Series: Journal of Physics: Conf. Series **919** (2017) 011003 doi:10.1088/1742-6596/919/1/011003

Organizing and Technical Committee

Alex. Sidorov,
Olesiy Nikitas,
Sergey Basalaev,
Artem Pomiluyko,
Artem Kuleshov,
Aleksandr Kozulin,
Yana Dubkova,
Ksenia Perfilieva

E-mail: cicmcm@niipmm.tsu.ru

IOP Conf. Series: Journal of Physics: Conf. Series **919** (2017) 011004 doi:10.1088/1742-6596/919/1/011004

Conference's Sponsor

Russia Foundation for Basic Research
Kuzbass Specialized Department for the production of drilling and blasting operations "KuzbasSpetsVzryv"
Tomsk Regional Public Organization "Association of Practical Shooting"
Academic book
Computer systems

IOP Conf. Series: Journal of Physics: Conf. Series **919** (2017) 011005 doi:10.1088/1742-6596/919/1/011005

Peer review statement

All papers published in this volume of *Journal of Physics: Conference Series* have been peer reviewed through processes administered by the proceedings Editors. Reviews were conducted by expert referees to the professional and scientific standards expected of a proceedings journal published by IOP Publishing.

IOP Conf. Series: Journal of Physics: Conf. Series **919** (2017) 012001 doi:10.1088/1742-6596/919/1/012001

Physical and mathematical modeling of pollutant emissions when burning peat

A Vasilyev[1], V Lozhkin[2], D Tarkhov[1], O Lozhkina[2] and V Timofeev[2]
[1] Peter the Great St. Petersburg Polytechnic University, 195251, Polytechnicheskaya, 29, St. Petersburg, Russia
[2] St. Petersburg University of State Fire Service of EMERCOM of Russia, 196105, Moskovsky, 149, St. Petersburg, Russia

E-mail: a.n.vasilyev@gmail.com

Abstract. The article presents an original neural network model of CO dispersion around the experimentally simulated peat fire. It is a self-learning model considering both the measured CO concentrations in the smoke cloud and the refined coefficients of the main equation. The method is recommended for the development of air quality control and forecasting systems.

1. Introduction

Peat deposits are found in many places, but the world's largest peatlands are the West Siberian Lowland, the Hudson Bay Lowland, and the Mackenzie River Valley [1]. Indonesia is known to have the largest tropical peatlands than any other country in the world [2]. The peat ecosystem is the most efficient carbon sink on the planet, for that reason, peat fires are significant sources of carbon dioxide and other greenhouse gases [3]. Peat fires produce a large amount of smoke that negatively affects air quality, public health, social activities, and the peatland ecosystem as a whole. In addition, peat fires release mercury into the atmosphere at a rate 15 times greater than upland forests, which may be a serious human health concern [4].

Peat land fires are of major concern in the Russian Federation [5-7]. The record-breaking heat wave and over 600 peatland and forest fires in the Central European part of Russia in summer 2010 resulted in a heavy smog blanketing large urban regions and caused numerous public health problems as a consequence [8].

There are two types of fires in peatlands: flaming fires and smouldering fires. The first ones occur at a temperature above 600 K and have high combustion efficiency, with more CO_2 and less smoke emitted. The second ones have higher CO and smoke emissions. Fires in peat tend to fall in the smouldering phase and persist in such a way for extended periods of time [9].

Peat can burn deep underground for meters, even in damp conditions and in winter time under the snow layer (Fig. 1).

Peat fires are difficult to extinguish, and severe fires in peatlands can last for months. Winter peat fires are often smoldering fires that create a lot of smoke because of incomplete combustion and result in greater emissions of carbon monoxide and other harmful substances including PM_{10} and $PM_{2.5}$, VOCs, etc. [5, 9-11].

If a peat fire develops near a motorway, the smoke from the burning peat-bog reduces the visibility, makes the breathing difficult, affects the human nervous and cardiovascular systems and may finally result in traffic accidents or in an emergency.

Such an emergency, caused by a peat fire, lasted in Irkutsk Region (situated in the West Siberian Lowland) near the Federal Highway "Siberia" from 26.10.15 to 15.01.16 (Fig. 2).

Figure 1. Smouldering peat bog near the Federal motorway P-255 "Siberia" in fall-winter 2015-2016

Figure 2. Emergency situation at the Federal motorway P-255 "Siberia" caused by a nearby peat fire in fall-winter 2015-2016

Atmospheric chemical transport models are widely used for the management of air quality strategies [11-13], however the accurate temporal and spatial estimates of emissions from peat fires are problematic because of uncertainties in the size, location, quantity of fuel consumed, relative proportions of flaming versus smoldering combustion, their temporal and spatial variability [9, 12]. We have previously developed an original neural network model on the base of real experimental data of CO emissions from the peat fire happened in the Irkutsk region near the Federal Motorway P-255 "Siberia" in fall-winter 2015-2016 [14, 15]. The results of numerical investigations showed that the proposed approach could be recommended as a useful tool for the air quality management. However, the technique still needs further improving through the obtaining of more accurate primary data that can be received from the experimental simulation of the peat fire, and through the step-by-step self-learning of the neural network model.

2. Modeling technique

2.1. Experimental procedure and the results of direct CO measurements

In order to develop a correct experiment simulating real peat fire, we should clarify the mechanism of fire development. Peat is a product of incomplete decomposition of plant materials under the conditions of excess humidity and insufficient aeration [6, 7]. The structure and composition can be modeled by a porous multiphase reactive medium consisting of cellulose, lignin, water in a liquid-droplet state, condensed products of pyrolysis, ash, and a gas phase [6, 7]. Peat has a high carbon content and can burn under low moisture conditions. Once ignited by the presence of a heat source (e.g., a burning match or a cigarette), it smolders for very long periods of time. Fig. 3 shows a schematic cross-section of the combustion zone of the surface peat layers [16].

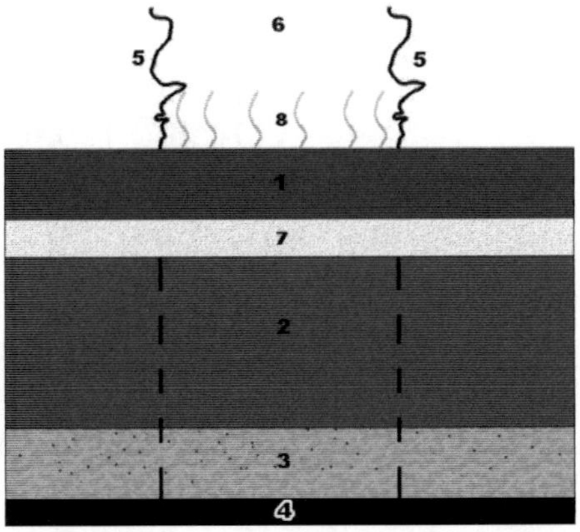

Figure 3. Schematic cross-section of the combustion zone of the surface peat: 1 is the soil layer, 2 is the peat layer, 3 is the water layer, 4 is the priming layer, 5 & 8 are the "smoke geysers", 7 is the snow (ice) layer, 6 is the smoke released from the burning peat

A peat fire was simulated outdoor using 25 kg of a natural peat sampled in the Leningrad Region. The sample of the peat was laid on a fireproof substrate and was burned using charcoal (Fig. 4). The peat has been smoldering in a flameless combustion mode for 40 minutes. The place of ignition was considered as zero in the measurement experiments (Fig. 5).

Figure 4. Simulation of a "smoke geyser"

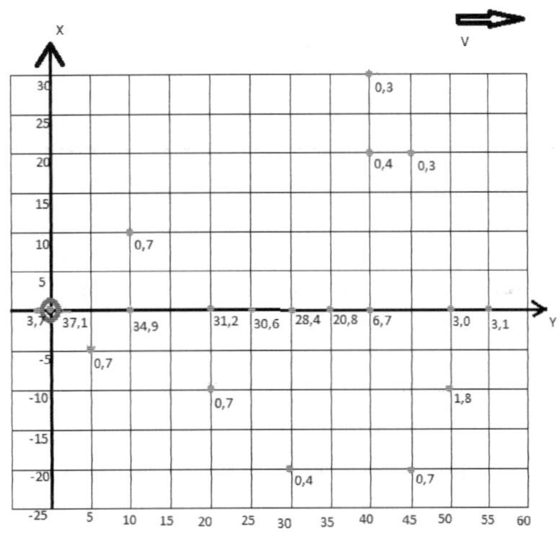

Figure 5. CO concentrations (blue points), mg/m^3, at different distance, m, from the simulated peat fire (red circle) considered as zero in the Cartesian coordinate system

We used two types of gas analyzers for the measurement of CO concentrations. Testo-330-2LL suits for the determination of CO content in combustion gases, and we measured with it the concentration of carbon monoxide near the simulated peat fire. PGA-200 is equipped with a more sensitive detector, and it is designed for the determination of ambient CO concentrations. The speed of wind was measured using the anemometer ATE-1033. The wind velocity during the experiment varied within 3-5 m/s. Technical characteristics of all instruments are summarized in Table 1.

Table 1. Technical characteristics of the measuring instruments

Measuring instrument	Measured Parameter	Detector	Detection limits	Accuracy
PGA-200 (Electrostandart,	CO	Electrochemical	0-120 mg/m^3	± 5 mg/m^3

Russia)				
Testo-330-2LL (Testo, Germany)	CO	Electrochemical	0-8000 ppm	± 10 ppm (0-200 ppm); ± 20 ppm (201-2000 ppm); ± 10% (2001-8000 ppm)
ATE-1033 (Aktakom, Russia)	Speed of wind	Remote impeller	0,4...30,0 m/s	± 2%

The results of the CO concentration measurements in the smoke propagation zone around the simulated peat fire at different distances from it are shown as blue points in the Fig. 5. We inputted these values as primary data when we developed the mathematical model of the pollutant diffusion in the air.

2.2. Justification of the mathematical model. Neural network approach

The equation of diffusion in the atmosphere, based on the physical law of conservation of the flow of matter. According to Berlyand [17-18], such parameters as instant concentrations of CO pulsed deviations from these values and the velocity of the CO diffusion should be taken into consideration while developing an emission model of the peat fire near the motorway. The problem is simplified by the application of the turbulent diffusion model:

$$\frac{\partial q}{\partial t}+u\frac{\partial q}{\partial x}+v\frac{\partial q}{\partial y}+w\frac{\partial q}{\partial z}=\frac{\partial}{\partial x}k_x\frac{\partial q}{\partial x}+\frac{\partial}{\partial y}k_y\frac{\partial q}{\partial y}+\frac{\partial}{\partial z}k_z\frac{\partial q}{\partial z}-\alpha q \ , \tag{1}$$

where q is the concentration of pollutant (*g/m*^{3}); x and y are the horizontal axis (*m*); z is the vertical axis (*m*); t is the time (*s*); u, v, w are the components of the wind speed in the direction of the axes x, y, z, respectively, (*m/s*); k_x, k_y, k_z are the components of the coefficient of turbulent diffusion along the Cartesian coordinate axes; α is the coefficient taking into account probable metabolism of CO in the atmosphere.

The authors of the present paper have a long-term positive experience of the application of this model for the estimation and the forecast of transport and fire-related air pollution (K-theory) [19-21]. The K-theory approach takes a long time and does not indicate inaccurate parameters of the problem, obtained from measurements. To solve these problems, we propose to apply the neural network approach.

On the base of the measurements, there was developed a neural network model with parameters (weights) tuned via optimization methods. The RProp-method and the combination of Cloud- and RProp-methods were in use [22-24]. The neural network model of the complex system can gather pieces of heterogeneous information – differential equations, conservation laws, equations of state, symmetry conditions, etc. The information exchange via neural network parameters between different levels of hierarchy makes computing less resource consuming.

Equation (1) can be greatly simplified for the solution to the problem concerning peat fire. In the steady state, we can assume that $\frac{\partial q}{\partial t}=0$.

The OZ – axis pointed upward, so, w for heavy particles, having own sedimentation rate (with the sign "$-$"), is equal to that rate, and w for light fractions is equal to 0. We also assume that the average turbulent flow of pollutants at the earth surface is small and the exchange coefficients are constant. We consider further only the light fraction and suppose that the concentration of CO satisfies the equation:

$$u\frac{\partial q}{\partial x}+v\frac{\partial q}{\partial y}-k\left(\frac{\partial^2 q}{\partial x^2}+\frac{\partial^2 q}{\partial y^2}\right)=Q, \quad (x,y)\in\Omega \, , \tag{2}$$

where Q is constant in the area of peat fire $K \subset \Omega$, ($Q = 0$ out of the fire); k is the known parameter, u, v are the known wind speed components corresponding to the measurements.

We can move the origin to the center of the peat fire, which we consider a superficial source of CO emission having a circular geometry.

In the second place, we orient the OX – axis in the direction of the wind by turning the coordinates. For the physical model of the "smoke geyser" we can assume that the component v of the wind speed in the direction of the OY – axis is equal to 0.

We deal further with such parameters as the radius of the area of the peat fire and the wind speed u. In the first version of the program, they are fixed and determined at the beginning as other input data. They can be changed further by the program learning. We will find a solution of the equation (2) in the form of neural network model:

$$q(x,y) = \sum_{i=1}^{N} c_i \exp\left\{\left[-a_i\left(x - x_i\right)^2 - b_i\left(y - y_i\right)^2\right]\right\}, \tag{3}$$

whose weights – parameters a_i, b_i, x_i, y_i and c_i – can be found by minimizing the error functional:

$$J = \sum_{j=1}^{M} \left| u\frac{\partial q(\xi_j,\eta_j)}{\partial x} - \frac{\partial^2 q(\xi_j,\eta_j)}{\partial x^2} - \frac{\partial^2 q(\xi_j,\eta_j)}{\partial y^2} - Q(\xi_j,\eta_j)\right|^2, \tag{4}$$

where $\left\{(\xi_j,\eta_j)\right\}_{j=1}^{M}$ are the trial points in the area where the solution is built, they are updated every 3-5 steps of the error functional minimization; $Q(\xi_j,\eta_j) = 1$ in the area of the peat fire K and $Q(\xi_j,\eta_j) = 0$ at other points out of the area.

We transform (2) to the form

$$\frac{\partial^2 q}{\partial x^2} + \frac{\partial^2 q}{\partial y^2} = \frac{u}{k}\frac{\partial q}{\partial x} - \frac{Q}{k}. \tag{5}$$

Thus, the solution to the problem is reduced to the development of an approach for determining the parameters $V_x = \frac{u}{k}$ and $\frac{Q}{k}$.

3. Results of mathematical neural network simulation of CO diffusion from the "smoke geyser"

We applied a three-step methodical approach to develop an adequate neural network model:

1) Construction of the neural network approximation using only the data from the experimental measurements;
2) Determination of the coefficients in the equation (2) using the approximation received at the first step;
3) Simultaneous determination of the equation coefficients and the construction of the neural network approximation using both experimental data (step 1) and the refined equation (step 2).

An appropriate computing program was developed for each step.

3.1. Step 1

Step 1 deals with the neural network approximation built on the base of experimental measurements. Fig. 6 shows the neural network approximation of CO dispersion at the height of human respiration (1.5 m) in the entire area of the smoke propagation.

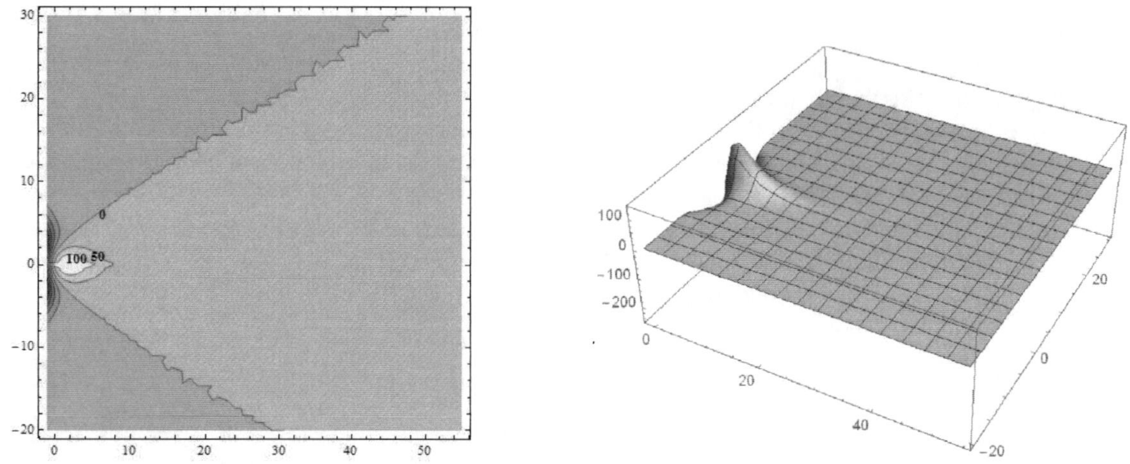

Figure 6. Step 1: Visualized approximation of the measured CO concentrations around the "smoke geyser" (5 particles in the cloud; 4 neurons)

3.2. Step 2

The results of Step 1 allowed us to determine the coefficients in the equation (2) or (5) and find the neural network solutions for the CO dispersion right above the smoldering source. We used random initial values of the coefficients in the equation (5). Fig. 7 demonstrates the neural network solution of the stationary process of CO diffusion directly above the burning source.

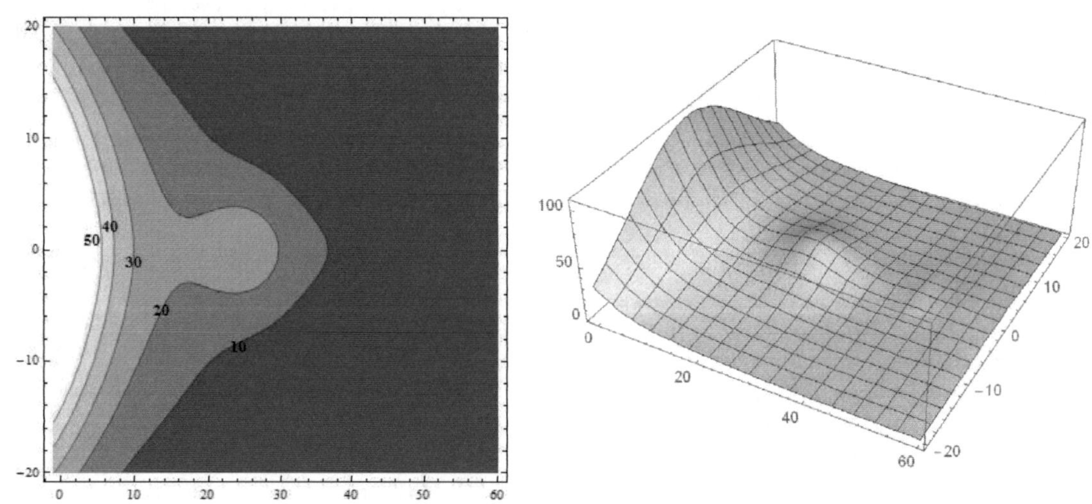

Figure 7. Step 2: Visualized approximation of the computed CO diffusion directly above the burning source (1 particle in the cloud; 150 neurons)

3.3. Step 3

Step 3 supposes implication of heterogeneous information, namely, the results of the real measurements from the Step 1 and the equation coefficients from Step 2, in order to find the neural network solution and to refine the coefficients of the equation (5).

The results of calculation were unstable for the random initial coefficients of the equation. In order to overcome this problem, we determined the coefficients of the equation (5) by a linear regression performed on the base of 20,000 random points through the substitution in the equation of the trial function realized in the Step 1 computing program. Such an approach allowed us to find fixed values of the coefficients: $Q = 1741.88$; $V_x = 13.73$, and run the neural network approximation of CO dispersion at the level of human respiration (1.5 m) in the entire area of smoke propagation in the case of fixed coefficients (Fig. 8) and in the case of fixed starting coefficients (Fig. 9) using both the results of the experimental measurements and the refined equation.

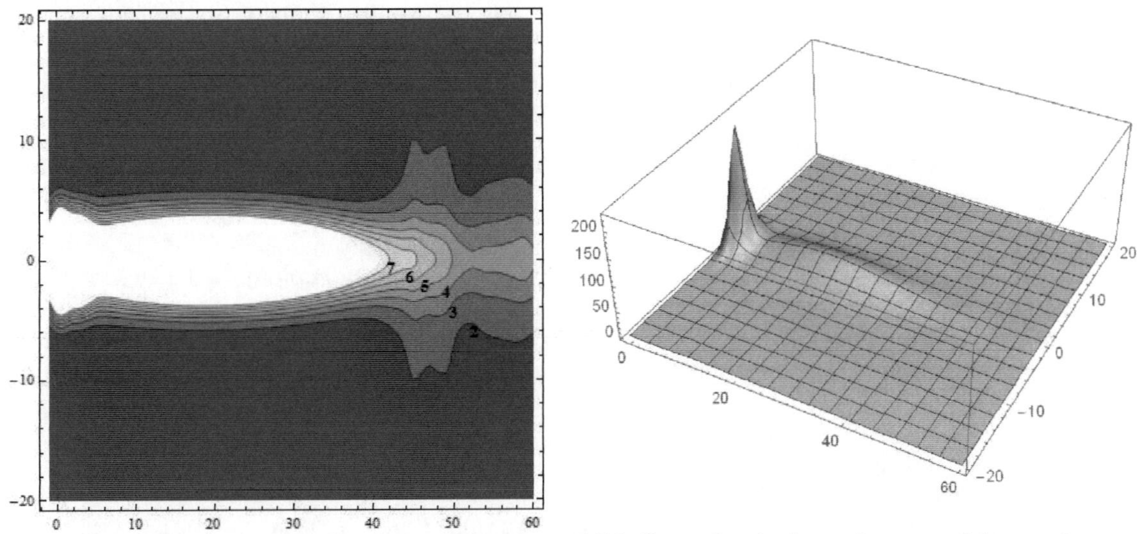

Figure 8. Step 3: Visualized approximation of CO dispersion in the entire area of the smoke propagation, received using the experimental data and the refined equation (fixed coefficients; 1 particle in the cloud; 50 neurons)

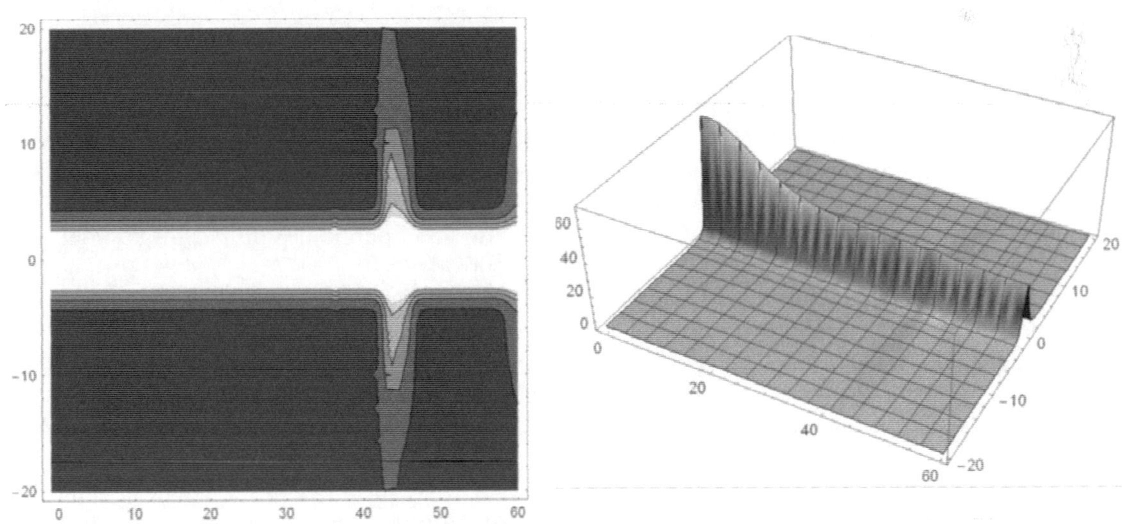

Figure 9. Step 3: Visualized approximation of CO dispersion in the entire area of the smoke propagation, received using the experimental data and the refined equation (fixed starting coefficients; 1 particle in the cloud; 150 neurons)

Conclusions

Some variants of approaches for constructing a neural network model of CO diffusion from a model source with a burning peat simulating a "smoke geyser" on a peat bog were developed and tested. The model is based on a numerical neural network solution of the atmospheric diffusion equation that realizes the fundamental physical law of matter conservation.

The availability of information on the "fire load" (mass of peat), the burning surface and the concentrations of CO directly above the combustion zone allows to calculate the values of CO, then, – the CO concentration field in the entire area of smoke propagation.

The method is recommended for the development of air quality control and forecasting systems.

The authors are going to devote a separate publication to methods of constructing neural network models for air pollution with other harmful components – the peat fire particles $PM_{2.5} - PM_{10}$.

References

[1] Fraser L H, Keddy P A 2005 The World's Largest Wetlands: Ecology and Conservation. Cambridge University Press, Cambridge, UK 488 p

[2] Tosca M G, Randerson J T, Zender C S, et al 2011 Dynamics of fire plumes and smoke clouds associated with peat and deforestation fires in Indonesia *Journal of Geophysical Research-Atmospheres* **116** 1-14

[3] Van der Werf G R, Randerson J T, Giglio L, Collatz G J, Mu M, Kasibhatla P S 2010 Global fire emissions and the contribution of deforestation, savanna, forest, agricultural, and peat fires (1997-2009) *Atmospheric Chemistry and Physics* **10** 11707-11735

[4] Biester H, Bindler R 2009 Modelling Past Mercury Deposition from Peat Bogs – The Influence of Peat Structure and ^{210}Pb Mobility *Working Papers of the Finnish Forest Research Institute **128*** p 483

[5] Fokeeva E V, Safronov A N, Rakitin V S, et al 2011 Investigation of the 2010 July–August fires impact on carbon monoxide atmospheric pollution in Moscow and its outskirts, estimating of emissions *Izv. Atmos. Ocean. Phys.* **47(6)** 682-698

[6] Grishin A M, Golovanov A N, Sukov Ya V, Abramovskikh A A. 2007 Experimental investigation of peat ignition and combustion regimes *Journal of Engineering Physics and Thermophysics* **80(6)** 1154-1157.

[7] Grishin A M, Golovanov A N, Sukov Ya V, Preis Yu I 2006 Experimental study of peat ignition and combustion *Journal of Engineering Physics and Thermophysics* **79(3)** 563-568

[8] Revich B A, Avaliani S L, Simons G 2016 Air pollution and public health in a megalopolis: a case study of Moscow Economy of Region 12(4) 1069-1078

[9] Elvidge C, Zhizhin M, et al 2015 Long-wave infrared identification of smoldering peat fires in Indonesia with nighttime Landsat data *Environ Res Lett* **10** doi:10.1088/1748-9326/10/6/065002

[10] De Groot W 2012 Peatland fires and carbon emissions. Natural Resources Canada. Canadian Forest Service. Great Lakes Forestry Centre, *Frontline Express (GLFC - Sault Ste. Marie)* **50** 2p.

[11] Konecny K, Ballhorn U, et al 2016 Variable carbon losses from recurrent fires in drained tropical peatlands *Global Change Biology* **22(4)** 1469-1480

[12] Gaveau D L A, Salim M A, et al 2014 Major atmospheric emissions from peat fires in Southeast Asia during non-drought years: evidence from the 2013 Sumatran fires *Sc. Rep.* **4** 6112

[13] Pouliot G, Pierce T, et al 2005 Wildfire Emission Modeling: Integrating BlueSky and SMOKE *Proc. 14th Int. Emission Inventory Conference "Transforming Emission Inventories - Meeting Future Challenges Today" (Las Vegas 11-14 April 2005)* Ed EPA US

[14] Lozhkin V, Tarkhov D, Timofeev V, Lozhkina O and Vasilyev A 2016 Differential neural

network approach in information process for prediction of roadside air pollution by peat fire *IOP Conference Series: Materials Science and Engineering* **V 158 (1)** doi:10.1088/1757-899X/158/1/012063

[15] Vasilyev A, Lozhkin V, Lozhkina O, Tarkhov D, Timofeev V 2016 Neural network approach in information process for predicting highway area air pollution by peat fire *CEUR Workshop Proceedings (CEUR-WS.org): Selected Papers of the XI International Scientific-Practical Conference Modern Information Technologies and IT-Education (SITITO 2016) (Moscow 25-26 November 2016)* **1761** 386-392.

[16] Usup A, Hashimoto Y, Takahashi H, and Hayasaka H 2004 Combustion and thermal characteristics of peat fire in tropical peatland in Central Kalimantan, Indonesia *Tropics* **14(1)** 1-19

[17] Berlyand M E 1991 Prediction and Regulation of Air Pollution 1991 *Atmospheric Sciences Library* **14** 320 p

[18] Genikhovich E L, Schiermeier F A 1995. Comparison of the United States and Russian complex terrain diffusion models developed for regulatory applications. *Atmos. Environ.* **29 (17)** 2375-2385.

[19] Ložkin V, Ložkina O, Ušakov A 2013 Using K-theory in geographic information investigations of critical-level pollution of atmosphere in the vicinity of motor roads *World Applied Science Journal* **23 (13)** 96-100

[20] Lozhkina O, Nevmerzhitsky N, Lozhkin V 2016 Evaluation of air pollution by PM10 and PM2.5 on St. Petersburg ring road: mobile measurements and source apportionment modelling *Proc. 10th Int. Conf. on Air Quality: Science and Application (Milano 14-18 March 2016)* Ed S Finardi, A Farrow et al (Hertfordshire: University of Hertfordshire) p 176

[21] Lozhkina O V, Lozhkin V N 2015 Estimation of road transport-related air pollution in Saint Petersburg using European and Russian calculation models *Transport. Res. Part D* **36** 178-189

[22] Tarkhov D, Vasilyev A 2005 New neural network technique to the numerical solution of mathematical physics problems. II: Complicated and nonstandard problems *Optical Memory and Neural Networks (Information Optics)* **14** 97-122

[23] Vasilyev A, Tarkhov D 2009 *Neural Network Modeling. Principles. Algorithms. Applications* (SPbSPU Publishing House, Saint-Petersburg) p 528 (in Russian)

[24] Vasilyev A, Tarkhov D 2014 Mathematical Models of Complex Systems on the Basis of Artificial Neural Networks *Nonlinear Phenomena in Complex Systems* **17** 327-35

IOP Conf. Series: Journal of Physics: Conf. Series **919** (2017) 012002 doi:10.1088/1742-6596/919/1/012002

Experimental study of the penetrating of plates by projectile at low initial speeds

M Yu Orlov[1], Yu N Orlova[2], Ig L Smakotin[1], V P Glazyrin[1] and Yu N Orlov[1]

[1]Tomsk State University, 36 Lenin Street, Tomsk, 634050 Russia
[2]Tomsk Polytechnic University, 30 Lenin Street, Tomsk, 634050 Russia
E-mail: orloff_m@mail.ru

Abstract. The research of the penetration process of lightweight plates by a projectile in the range of initial velocities up to 325 m/s was attempted. The projectile was a shell bullet and the barriers were of ice, MDF-panels and plexiglas barriers. The response of barriers to impact loading is studied. High-speed shooting of each experiment is obtained, including photos of the front and rear sides of the barriers. An attempt was made to reproduce the scenario of the destruction of barriers. The results of experiments can be interpreted only as qualitative tests. Projectile was not destroyed.

1. Introduction

The processes occurring in solids under high-speed deformation are now the subject of fundamental and applied research, both in Russian Federation and abroad. This is due, above all, the breadth of application use in various spheres of human life. This includes: the creation of effective anti-shock protection of civilian and military facilities and equipment; welding and explosion cutting; hydrostretching; shock-wave pressing; explosive hardening; safety of the shells of a nuclear reactor in cases of falling into them objects from outside (aircraft, fragments, etc.) or loading from the inside (excessively high pressures that arise when the reactor is broken), etc. Separately, one can single out the protection of space vehicles from the threat of micrometeorites and technogenic debris particles impacting them. In addition, the accumulated experience in the field of high-speed deformation of solids is of interest in medicine, astrophysics and military applications [1-5].

The interaction of solids is accompanied by complex processes, the final role of which is determined by the presence of a number of factors: the initial velocity of the impacting objects, shape, and physics and mechanical property. It should be noted that in most cases, during an interaction, penetration of the bodies into each other takes place, as well as the melting and destruction of the material into separate fragments.

In the quantitative description of high-speed shock phenomena, many complex problems arise, which at present are far from solving and require the use of mathematical modeling and experimental methods of investigation. Certain results can be obtained by carrying out large-scale model tests and full-scale tests using various ballistic set up and explosive accelerators for throwing bodies. For example, the size of the fragment, the level of its destruction, the shape and depth of the crater, the ballistic limit, etc. However, it is necessary to note the technical complexity and high cost of carrying out such experiments, as well as the impossibility of obtaining detailed information on the space-time distribution of stress fields, deformations and fracture regions. Nevertheless, the importance of obtaining experimental results does not cause doubts.

The laboratory 21 National Research Tomsk State University developed a multifunctional user program complex for solving modern multi-contact dynamic problems of solid body mechanics. Under such problems, one should understand the problems of the penetration and deep penetration by the impactor of complex geometry into structurally heterogeneous barriers. Several solvers have been developed for two-dimensional plane and axisymmetric statements [6]. The software package is developed as an alternative to commercial FEM software LS-Dyna.

A numerical lagrangian method is developed for calculating modern dynamic multi-contact problems of mechanics of a deformed solid. The originality of the method lies in a new way for isolating the surfaces of discontinuity in the continuity of materials, which does not impose serious limitations on the solution of these problems. The numerical method contains an algorithm for splitting nodes, erosion triangulated elements, rearranging the contact and free surfaces, and so on [7].

The latest innovation is the mobile laboratory "Explosive Destruction of Natural Materials", organized in Tomsk State University. The mobilab expands the research capabilities of the team, and its main goal is full-scale experiments. The one is functioning with the support of the EMERCOM of Russia and KuzbasSpetsVzryv company. Full-scale tests on underwater explosion of ice and explosion of limestone are carried out. The diameters of blasting lanes in the ice and the shape of the explosion crater in limestone were established [8].

In the current research, the behavior of freshwater ice, MDF-panels and plexiglas at impact load are investigated. A snap-analysis of the behavior of these materials under impact loading was carried out. High-speed shooting of the penetration process is obtained. Studied in detail the inlet and outlet openings into the barriers. An attempt has been made to study the morphology of destruction. In all cases, the form of the projectile after penetration is established.

2. Experimental set-up

This section focus on the experimental set-up. The experimental set-up was installed in the Tomsk Field Shooting Society is shown in figure 1. This experimental setup is mobile and designed specifically for snap analysis of the penetration process. The main components were a bullet receiver, light projectors, a special place for targets and a high-speed camera.

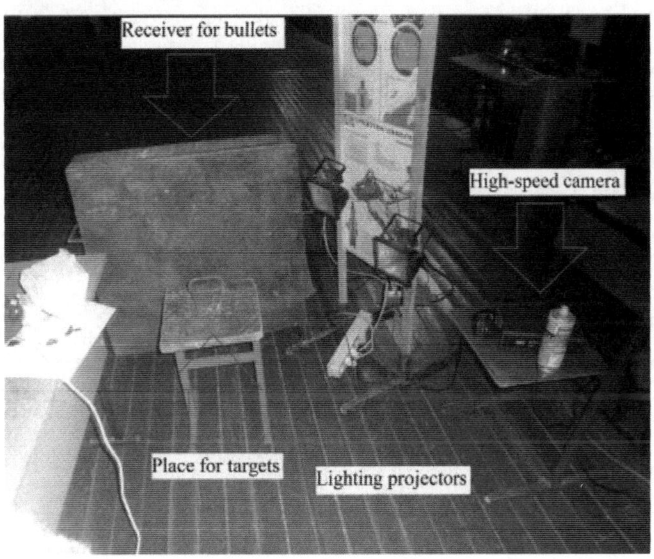

Figure 1. The experimental set-up. Photo by M. Orlov

The presence of the bullet receiver excludes a detailed study of the impactor after the shock. The one present only in the first experiment, because in this case the form of the impactor was known.

Several light projectors were used. The research objects can be moved independently in the different degrees of freedom.

A high-speed camera (Photron Fastcam APX RS, 10000 frames per second) was used to film the impact loading. There were several camera positions around the barrier. The first position was perpendicular to the barrier is shown in figure 1. The remaining positions were at an obtuse angle and an acute angle to the barrier. This allowed for a more detailed study of the process of penetration.

2.1. Projectile and initial velocity

Only one projectile was used. The projectile is a shell bullet. Weight is 6.9 g. Length is 12 mm. The diameter of the leading part is 9.2 mm. Shell weight is 1.5 g. The shell material is low-alloy steel. Core weight is 2.5 g. The material of the core is plumbum. Initial velocity is 325 m/s. The projectile is depicted in figure 6.

3. Research ice penetration by projectile

The behavior of this ice under impact loads was investigated in [9-11]. The diameter of the crater and the depth of the crater in the ice blocks are revealed after penetrating by the spherical impactor. Earlier, it was established that the impact craters in the ice had a conical shape. The original research of penetration of the ice barriers is presented. One of the main objectives was to establish the main mechanisms of the process of destruction of ice barriers. The results of the interaction of these materials are predictable.

3.1. Sample preparation

The first object of research was freshwater ice. In a laboratory experiment were single-layer, dual-layer and three-layer ice barriers. However, current study single-layered barriers were considered only. The ice was obtained under special conditions: the freezing point was -17 ℃, the freezing time was ~ 36 hours. Prepared samples were placed in a special container and delivered to the laboratory. 25 samples were prepared. The diameter of each sample was 14.5 cm and the height of the sample was 4.5 cm. This ice was called congelation ice. The physical and mechanical properties of such ice are given in [12].

3.2. Projectile penetration analysis

The presented results are a logical continuation of systematic studies of the behavior of ice under dynamic loads. The following regularities were established in the experiment. After the penetration of the ice barrier, the projectile was not deformed. Most likely that the projectile velocity the barrier and after the barrier was approximately the same. The number of fragments separated from the front of the barrier and from the rear part of the barrier was also approximately the same. This is seen in figure 2. Figure 3 shows the barrier before the impact, and figure 4 shows the barrier after the impact.

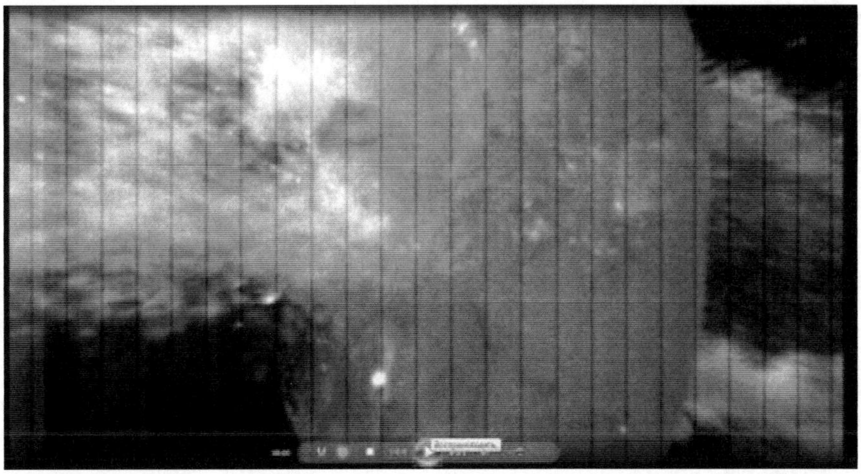

Figure 2. The penetration of the ice barrier by projectile

Figure 4 clearly shows four large fragments of ice. These fragments took place in high-speed photography. It is likely that these fragments were formed as a result of the propagation of the main cracks. Propagation direction was from the center to the side surfaces of the ice cylinder. Smaller fragments of ice are noted in the photograph. Unfortunately the projectile is not visible on this photo. Most likely, the distance between the projectile and the barrier is greater than the diameter of the ice barrier.

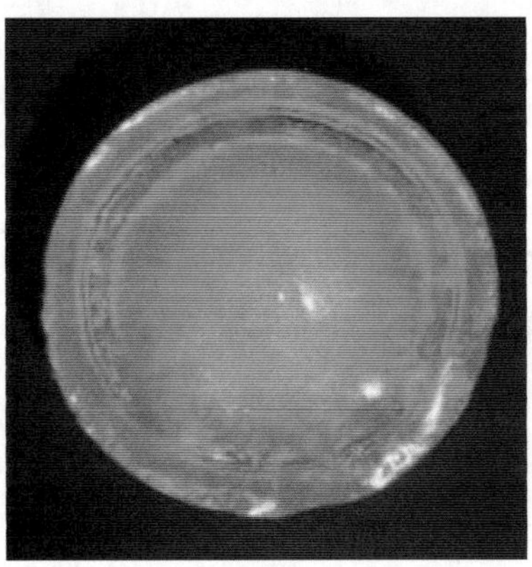

Figure 3. Ice barrier before impact

Figure 4. Ice barrier after impact. Scale bar represent 1 mm

Thus, the ice cylinder practically did not resist steel projectile penetration. The ice breaks up into fragments of various sizes, include very small ones (approximately 1 mm). Four fragments of ice (3 cm in size and slightly more) are clearly visible. Projectile after impact remained undamaged. Obviously this is due to the strength properties of the interacting materials.

4. Research MDF-panel penetration by projectile

The MDF-panels as the object of research were selected by the authors for the first time. The use MDF-panels in modern protective structures is barely possible. More likely, the use of MDF-panels and similar materials is possible as back layers. For example, the MDF-panel can be used as the back of armored door in banks, etc. Current experiment focuses on the expansion of scientific knowledge about the behavior of these materials under impact.

4.1. Research object

The next objects of research were the barriers of MDF-panels. These studies were planned to expand the scientific knowledge about the behavior of this material under shock loads. MDF panels are stronger than congelation ice.

Some mechanical properties are as follows:
The initial density is 0.95 g/cm^3.
Bending strength is 22 MPa.
Elastic modulus is 1,7 GPa.
The thickness of the barrier is 9 mm.
The diameter of the barrier is 200 mm.
The number of barriers in the tests is 10 pcs.

4.2. Projectile penetration analysis

In this paragraph, the results of penetration of the MDF plate are presented. It should be noted that high-speed shooting of this process was carried out at different angles to the barrier. This is necessary to fix the cloud of the leading fragments.

Figure 4 illustrates the process of penetrating a barrier by a projectile. Approximate time is 5 microseconds. It was revealed that the penetration was accompanied by the separation of fragments from the front and back sides of the barrier. Debris of broken away from the front of the barrier moved opposite to the movement of the projectile. The vast majority of the fragments were not large (less than a projectile). Only a few fragments measured 1 cm or more.

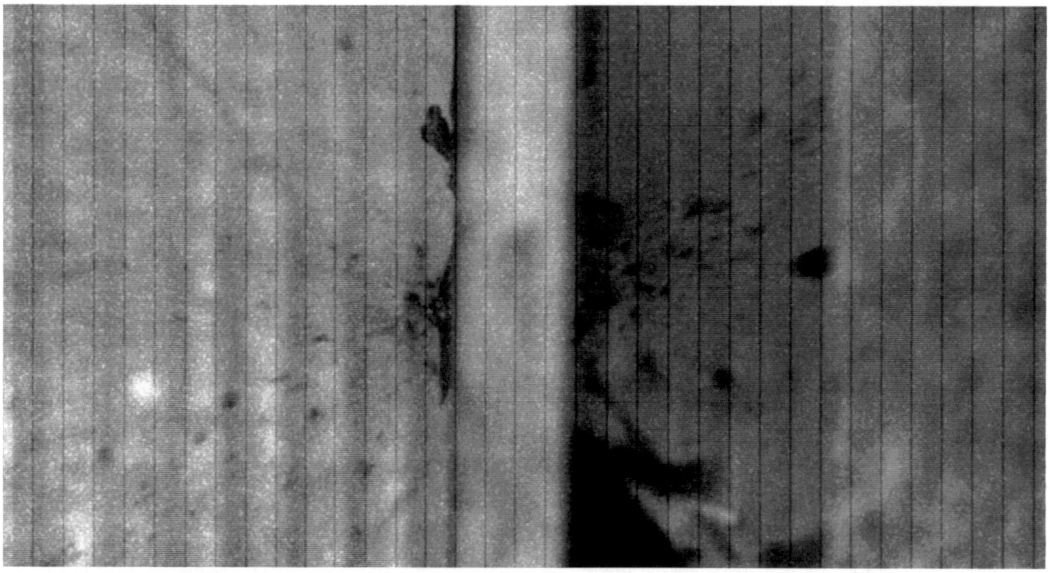

Figure 5. The penetration of the MDF-barrier by projectile

A distinctive feature should be considered the formation of a through hole in the barrier. The diameter of the hole was equal to the diameter of the projectile. And the internal surface of the hole was not damaged. The number of fragments separated from the rear side is more than the number of fragments from the front side MDF-barrier. The state of the projectile after the impact was the same as that subsection 3. However, there were scratches in its cylindrical part.

Figure 5 shows the obverse of the barrier from the MDF-panel. Here the entrance hole is clearly visible. The diameter of the hole was 9 mm, i.e. the diameter of the hole and the diameter of the projectile were the same. There are no annular and radial cracks on the surface.

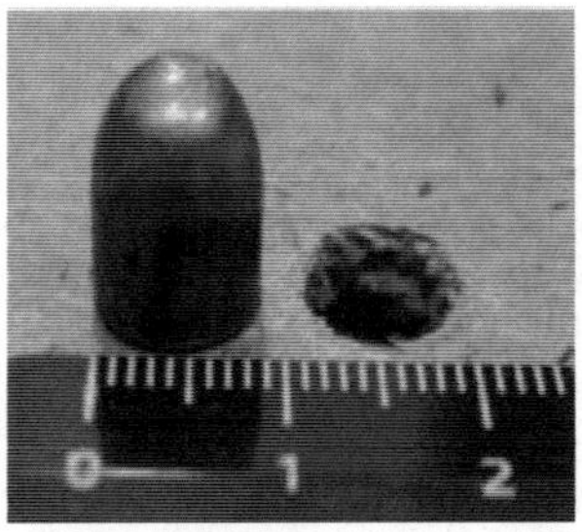

Figure 6. Front side barrier

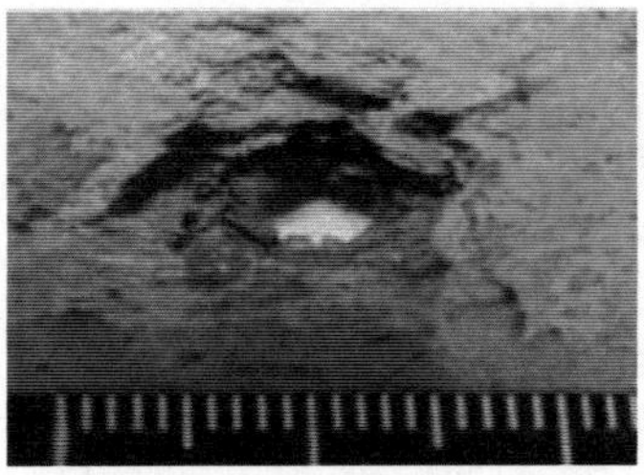

Figure 7. Rear side barrier

Figure 6 shows the back side barrier. It is seen that the barrier is more damaged in the rear. There is no near-surface layer 18 mm in diameter. The thickness of the layer is 2.5 mm. The photo shows holes of the specified diameter. There are no cracks on the back side. The research suggests the morphology of destruction the ice and MDF-panel after penetrating were different.

5. Research plexiglas penetration by projectile

The last object of research was organic glass (plexiglas). The choice of plexiglas as an object of research is due to the fact that it is a frost-resistant material and a dielectric. Understanding the dynamic response of plexigalas has important applications on engineering, including some protective structure and aircraft elements.

5.1. Research object

In research [13], the behavior of plexiglas was investigated in the speed range from 500 m/s to 5000 m/s. The behavior of plexiglas at such speeds was predicted, including penetration at all stages, hydrostatic pressure isolines and hydrostatic pressure at control points, shortening of the projectile, speed of free surface, penetrating time, the time of formation of the first foci of destruction, etc. The present research are limited only to the experiment of penetrating the plexiglas barrier by the projectile.

The initial density is 1.18 g/cm^3.

Tensile elastic modulus is 3 GPa.

Tensile strength is 70 MPa.

The thickness of the barrier is 10 mm.

The diameter of the barrier is 100 mm.

The number of barriers in the tests is 10 pcs.

5.2. Projectile penetration analysis

Figure 8 illustrates the process of penetrating barrier of plexiglas. As in figures 2, 5 the final stage of the process is shown. A few microseconds ago, the projectile penetrated and destroyed the barrier. It can be seen that the process of destruction of plexiglas differed from the processes of destruction of the MDF-panel and freshwater ice. The cloud of the leading fragments consists of small (about 1 mm) and medium (from 5 mm to 10 mm) fragments. An important point is that this cloud is shaped by the near-surface layers on the back of the barrier. For some time, the projectile is in contact with these fragments. After penetration, the barrier is destroyed by 6 or 7 large fragments (15 mm or more). The form of the projectile after penetration was the same as in the previous case.

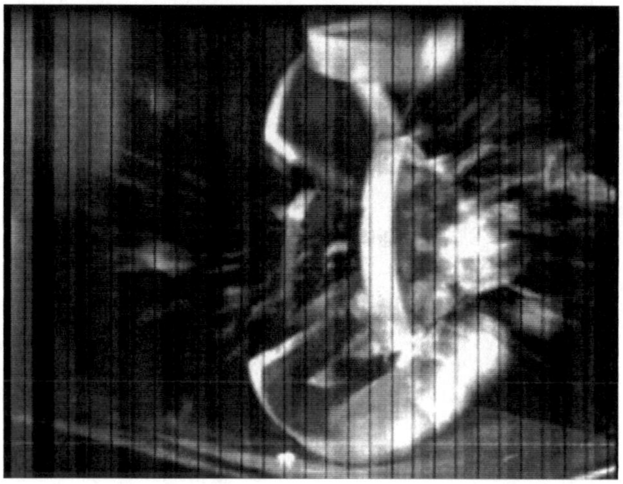

Figure 8. The penetration of the Plexiglas by projectile.

Figure 9 shows the rear side of the plexiglas barrier after the impact. Near the barrier are the fragments of medium size. Most of the fragments previously were back side of the barrier. In the center of the barrier is clearly visible hole. The diameter of this hole is approximately 14 mm. This zone consists of cracks inside the barrier. A significant part of the material from the rear side was beyond the barrier. The front side of the barrier also lost some volume of material. It is visually established that the lost volume of material from the front side was less than the lost volume from the back side of the barrier. From the center of the barrier, the main cracks spread to the lateral surfaces. In the future, according to these cracks, plexiglas destroyed. In research [14] such cracks were called radial.

Figure 9. The rear side of the barriers

Summary

The process of destruction of ice has been studied in detail. The ice was destroyed by various fragments, and there was no precise estimate of the ice fragments. A similar picture of the destruction can be observed in [15]. The projectile did not failed after the penetrating of the barrier. The velocity of the projectile after penetration has not been measured, but there is every reason to believe that it has not changed significantly. The result can be spent as a new qualitative test only.

The process of penetration of the MDF-panel by the projectile is also investigated. In contrast to the previous case, a through hole with a diameter of 9 mm was in the barrier. The destruction of the obverse side of the barrier and the destruction of the rear side of the barrier were different. On the front side there were no damage to the near-surface layer. No annular and radial cracks were observed, extending from the hole to the lateral surfaces of the barrier. After the penetration, the impact did not change its original form.

Later in the current research, the process of destruction of plexiglas was studied. Here, other features of the process of destruction were noted. The quality of the destruction of plexiglas is similar to the destruction of ice. However, in the first case there was a through hole. In this case, the number of medium fragments and large fragments were maximum. After penetration the projectile is not destroyed.

Conclusion

Thus, the process of penetrating barriers by a projectile with an ogive head part at a low initial speed has been studied. In all cases, the barriers were penetrated. In addition, plexiglas and ice were destroyed in fragments. This was explained by the size of barriers and their mechanical properties. In the MDF-panel there was a through hole only. The projectile was not destroyed after the impact.

Acknowledgments
The authors are grateful to A. Lisitsyn, E. Borzenko, A. Usanina, V. Solonenko, A. Yashchuk for their assistance in carrying out the research.

References

[1] Tria D, Ali B and Karim H 2017 Modeling assessment of ceramic\composite armour under ballistic impact *Proc. Int. Conf. on Mechanics and Materials in Design (Albufeira)* (Portugal) pp 397-399

[2] Crawford J, Choi H and Brewer Tim 2016 Simplified physics-based modeling projectile penetration analysis problems *Proc. 4rd Int. Conf. on Protective Structure (Beijing)* (China) pp 538-546

[3] Iqbal M A et al 2016 Effect of target span and boundary conditions on ballistic limit of thin aluminum plate *Proc. 3rd Int. Conf. on Protective Structure (Newcastle)* (Australia) pp 229-233

[4] Kim S B, Kim H W and Yoo 2015Penetration analysis of projectile with inclined concrete target EPJ Web of conference 94 04052

[5] Gilson L, Gallant J and Rabet L 2013 Study of a combined blast and ballistic protection. An experimental and numerical approach with three material models *Proc Int. Conf. 7th Symposium of Light Weight Armour Group (Brussels)* (Belgium) pp 17-25

[6] Orlov Yu N et al 2010 Impact. 2D. Shock-wave loading of structures. Axisymmetric problem (Certificate of state registration of the computer program 2010610911)

[7] Gerasimov A V et al 2007 *Theoretical and experimental studies of high-speed interaction of bodies* (Tomsk: Tomsk State University Press)

[8] Orlov M Yu, Orlova Yu N and Bogomolov G N 2016 The behavior limestone under explosive load *J. Phys. Conf. Series* **774** 012052

[9] Pernaz-Sanchez J et al 2016 Experimental analysis of ice sphere impacts on unidirectional carbon/epoxy laminates *International Journal of Impact Enginereeng* 96 1-10

[10] Combescure A et al 2011 Experimental study of high velocity impact and fracture of ice *International Journal of Solids and Structure* 48 2779-2790

[11] Gareth J et al 2011 On the response of commercially-important CFRC structure to multiply ice impacts *Composite Structure* 37 177-184

[12] Bogorodsky V V and Gavrilo V P 1980 *Ice. Physical properties. Modern methods of glaciology* (Leningrad, Publishing House Gidrometeoizdat) p 383

[13] Glazyrin V 2008 Deformation and destruction of inhomogeneous materials and structures in shock and explosion Ph.D. Thesis Tomsk, Russia

[14] Orlov M and Bogomolov G 2016 Study of the behavior of ice under shock loads *Polar mechanics pp* 192-202

[15] Pernas-Sanchez J, Arterro-Guerrero J, Varas D and Lopez-Puento J 201 Experimental and numerical analysis of ice sphere impact 7th *International Workshop on Dynamic behavoir of materials and its application in industrial processes-(Madrid)* Spain pp 47-49

IOP Conf. Series: Journal of Physics: Conf. Series **919** (2017) 012003 doi:10.1088/1742-6596/919/1/012003

Resistance of GPS samples to non-contact underwater explosion

A I Dulnev and E A Nekliudova

Krylov State Research Centre, Moskovskoe shosse 44, St. Petersburg 196158, Russia

E-mail: a_dulnev@ksrc.ru

Abstract. Test data obtained from experiments on GRP test samples in the explosion chamber of Krylov State Research Centre are given. Characteristic features of material damage accumulation and failure processes are discussed based on the experimental data. The sample material damage is assessed versus three indicative effects (criteria): binder failure, fiber rupture, through-thickness penetration (hole). Finite element modelling of the experiments is used for a more detailed analysis of stress-strain state of material samples.

Introduction

Materials of ship hull structures are subject to various kinds of loads during ship operation, including high-power dynamic loads induced by underwater explosions. In the recent years, polymer composite materials (PCM) have been increasingly used for manufacturing ship hull structures. In addition to their high strength-to-weight ratios, corrosion resistance, low heat conductivity and non-magnetic qualities these materials offer a wide choice of options to design robust structures capable to withstand both static and dynamic loads. Resistance of polymer composites to loads strongly depends on the reinforcement materials, binders, method of reinforcement, manufacturing processes, etc. PCM resistance has been widely discussed primarily in the context of quasi-static loading, while much less attention is paid to explosion-induced effects, in particular, underwater explosions (UNDEX). However, such investigations should be of interest in identification of the best PCM options for ship structures.

Various aspects of PCM responses to underwater explosions were examined in [1-12]. Analyses of PCM responses essentially rely on experimental data, and for this reason the test procedures and technologies are discussed at length in these references. In [3, 5, 6] PCM samples were loaded in a conical shock tube where micro charges were exploded to generate a plane shock wave. In [4, 8, 9] the tests were performed using a cylinder shock tube. In this case the UNDEX shock wave is generated when a heavy piston plate projected by a gas gun impacts and decelerates in water. It should be noted that in case of shock tubes the effective field of exposure to shock wave has a limited diameter of 230-250 mm, which imposes some constraints on the thickness, character of stress-strain state and types of test sample failure. The tests described in [10, 11] were conducted in an explosion chamber, test samples were mounted using special-purpose fixtures and exposed to shock waves generated by explosive charges of various masses. The tests covered in [1, 7] were performed in an open-water test range. In [1, 3-11] flat samples were tested. In [12] the results of internal explosion tests on cylinder PCM tubes are analyzed.

Along with experimental investigations, analytical methods and computer models of sample responses to UNDEX shock wave were elaborated in [2-6, 8-11], which enabled researchers, on the one hand, to verify the results of estimations and, on the other hand, to obtain additional insights from test data.

The main purposes of this study were

– to obtain test data on explosion resistance of GRP test samples exposed to a non-contact UNDEX, including evaluation of the extent and character of damage versus explosive loading parameters;

– to elaborate computer models and use computer simulation techniques for analysis of the stress-strain state of test samples as applied to given test conditions.

Tested material and samples

Test samples were manufactured from quadraxial glass fiber (stacking 0°/+45°/90°/-45°) and vinyl ester binder using the vacuum infusion process. Four sample batches were made and tested. Each batch contained samples of approximately uniform characteristics. Batches were different in thickness and, therefore, surface mass characteristics of samples. Characteristics of the samples are given in Table 1. The main mechanical characteristics of the composite material under study, which were obtained from static sample tests, are given in Table 2 [11].

Table 1. Main parameters of samples

Sample batch number	Number of samples	Thickness δ, mm	Density ρ, kg/m^3	Surface mass m, kg/m^2	Number of woven layers
1	6	7.91-8.75	1740-1860	14.71-15.23	10
2	5	7.36-7.40	1960-1990	14.49-14.69	10
3	5	5.18-5.20	1970-2000	10.20-10.40	7
4	5	3.66-3.75	1990	7.28-7.46	5

Table 2. Main mechanical characteristics of GRP

Characteristic	Testing direction, deg.	Symbol	Value
Modulus of elasticity, GPa	0	E_1	21.8
	45	E_{12}	20.8
	90	E_2	22.2
Modulus of shear in reinforcement plane, GPa	0	G_{12}	8.0
	45	G_{121}	8.9
	90	G_{21}	8.4
Modulus of interlaminar shear, GPa	0	G_{13}	4.2
	90	G_{23}	4.5
Poisson ratio	0	ν_{12}	0.29
Ultimate tensile elongation, %	0	ε_1	2.6
	90	ε_2	2.8
Ultimate tensile/compressive strength, MPa	0	σ_{11}	394/371
	45	σ_{1122}	334/368
	90	σ_{22}	410/357

The samples were circular plates of 300 to 400 mm radius. Each plate had 18 holes for bolts. Bolt setting radius was 250 mm. Samples of batches 3 and 4 were made with a circular mould-on portion in way of the bolt joints to achieve identical failure patterns for samples of different batches, i.e. to obtain damage (fiber rupture, penetration hole) in the middle of the sample and avoid any fractures at supports or near bolt joints, which are treated as accidental and largely dependent on the sample fixing arrangements. Total thickness at the mould-on portion was 14 mm.

IOP Conf. Series: Journal of Physics: Conf. Series **919** (2017) 012003 doi:10.1088/1742-6596/919/1/012003

Experimental set-up

The samples were tested in the explosion chamber of Krylov State Research Centre [10, 11]. The schematic arrangement of this facility is shown in figure 1a. The tests were performed using a so-called water-air set-up when the sample facing the explosion was backed by air-filled space. Each sample was exposed to one explosion. Every time an explosive charge was placed vis-a-vis the sample center at a fixed distance (300 mm), the charge mass Q was varied (from 8 to 90 g). The charge itself was a plastic explosive of cylinder shape with a height to diameter ratio of 1. The explosive composition was as follows: hexogen – 80 %, deterrent – 20 %. The explosive density was 1350 kg/cm, detonation velocity – 7200 m/s, TNT equivalent in terms of the specific blast energy was ~ 1. According to the relationships of [13] the shock wave acting on the test sample had the following parameters: maximum wave front pressure p_m=35…85 MPa, exponential decay constant θ=(2.0…3.1)·10^{-2} ms.

Figure 1b shows the test dummy assembly. The sample was fixed with 18 bolts between the matrix foundation (tubing) and clamping metal strip (ring). A rubber gasket was fitted between the clamp and the sample to seal the assembly. The effective test field diameter of the matrix-mounted sample was 400 mm. The test assembly was suspended vertically in the explosion chamber, which was filled with water so that the water surface was at least 1.0 m above the top of the assembly.

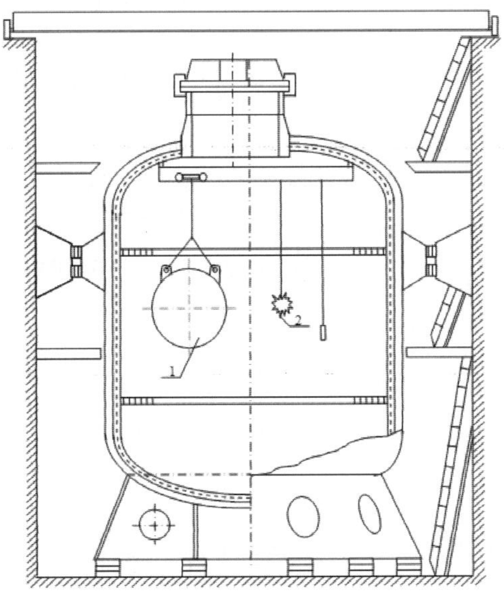

a) Explosion chamber:
1 – dummy, 2 – charge

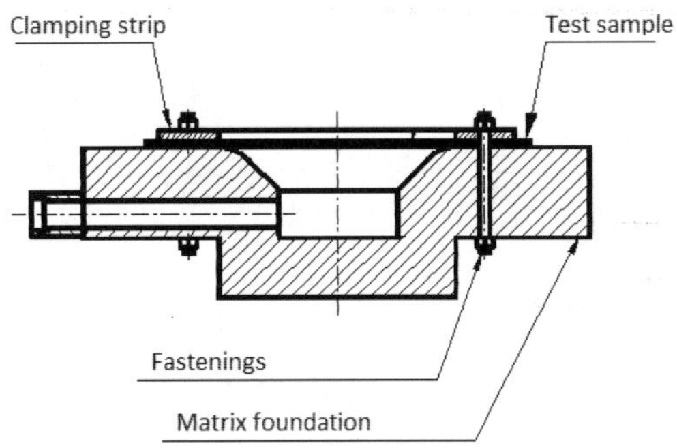

b) Dummy assembly

Figure 1. Explosion chamber and dummy assembly

During the tests the strains at the back side of the sample (i.e. on the opposite side to the explosion) and blast pressure in open water were recorded. Strain gauges to record radial and circular strains were fitted at different points around the sample center over a 125 mm radius circle. The arrangement of strain gauges is shown in figure 2. Pressure readings were used to monitor the completeness of charge detonation and TNT number as well as to check that the sample strains are not affected by any wave reflections from the chamber walls.

IOP Conf. Series: Journal of Physics: Conf. Series **919** (2017) 012003 doi:10.1088/1742-6596/919/1/012003

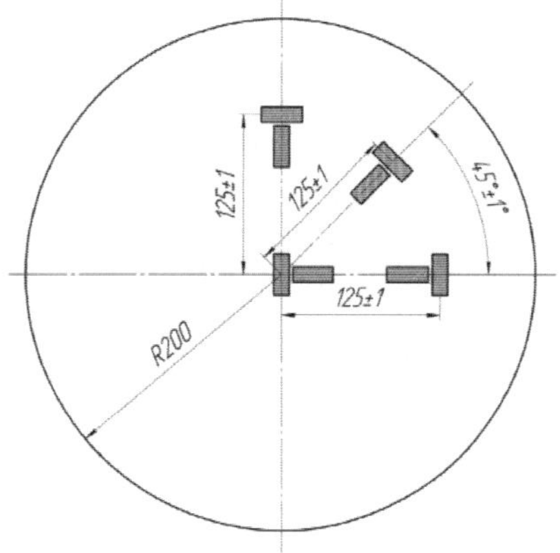

Figure 2. Arrangement of the strain gauges on the sample

Based on the test data, different levels of damage inflicted on samples in each test batch were correlated with the mass of explosive charges. The level of damage was assessed versus three characteristic effects (criteria): 1 – binder failure (visually identified after tests as a change in colour (bleaching) in the middle part of the sample; 2 – rupture of individual fibers; 3 – through-thickness penetration (hole). The maximum relative error in estimation of the charge mass for the damage by criteria 2 and 3 was within 15 %. In case of criterion 3 the relative error was somewhat higher, reaching 30 %, which is explained by small charge masses and test data scatter for binder failure as well as limited number of test samples.

In accordance with the above criteria three levels of explosion resistance were established for the samples. To generalize the test results obtained for the samples with different surface masses, a measure of the explosion effect was introduced: relative charge mass $\beta = \frac{m_{ex}}{m}$, where m_{ex} is the mass of charge per unit of the sample's test area, kg/m^2; m – surface mass of the sample, kg/m^2. Thus, the relative charge mass triggering one or another kind of failure is taken as a measure of specific explosion resistance of the samples.

Discussion of test results

Figure 3 shows photos of typical failure patterns of samples, and Table 3 contains the relative masses of charges characterizing different explosion resistance levels of samples.

a) Binder failure, back side (Q = 8 g, β = 0.44%, δ = 7.32 mm, m = 14.49 kg/m^2)

b) Fiber rupture in three layers, back side (Q = 70 g, β = 3.83%, δ = 7.40 mm, m = 14.50 kg/m^2)

c) Hole, front side (Q = 55 g, β = 4.20%, δ = 5.20 mm, m = 10.40 kg/m^2)

Figure 3. Typical damage patterns of samples

Table 3. Explosion resistance of test sample batches

Sample batch number	Surface mass, kg/m^2	Relative charge mass β for various explosion resistance levels,%		
		β_1	β_2	β_3
1	14.71-15.23	≈0.52	≈3.13	≈4.38
2	14.49-14.69	≤0.44	≈3.27	>3.83
3	10.20-10.40	≤0.62	≈3.50	≈3.85
4	7.28-7.46	≤0.87	≈3.77	–

As the explosion power (charge mass) is progressively increased it is seen that binders of test samples start to fail, first in the center and in the supported portion of a sample, then this damage spreads over the entire test area of the sample (figure 3a). It follows by rupture of individual glass fibers in the middle part, starting from the back layer. At this stage the fiber rupture is mostly chaotic (figure 3b). Finally a through-thickness penetration hole is formed by two main intersecting fractures propagating at about ± 45° to the reinforcement fiber material (figure 3c). The through-thickness penetration failure refers to approximately the same specific explosion resistance across all test batches of samples with a difference of less than 15 %, which practically falls within the test data error. It should be noted that samples of less thickness demonstrate a somewhat higher explosion resistance in terms of this criterion, apparently because of lower bending strains. However this difference is small, e.g., if the sample thickness is doubled, β_2 changes only within 20 %.

The energy absorption capacity of samples before initiation of fiber ruptures is mainly governed by the binder failure. Thus, after initiation of binder failure the relative charge mass has to increase more than 5 times to cause a through-thickness penetration, while after initiation of fiber rupture only a 10 to 40 % increase is sufficient to initiate this type of failure. Two comparative tests with batch 2 samples were performed to confirm the effect of binder failure on the energy absorption capacity of samples. In one case the sample was first subjected to a 8 g charge explosion (β=0.44 %). Here, the binder failure occurred (from the back side throughout practically the entire test area, from the face side in the area of the support contour) but no fiber ruptures were observed. Then the same test sample was subjected to a 70 g charge explosion (β=3.83 %), which resulted in the hole. In the second case, the similar sample was subject to a single explosion of a 70 g charge (β=3.83 %). Here, the explosion produced no hole, resulting only in fiber ruptures from the back side at the center of the sample. Thus, preliminary shock effect on the sample (causing a binder failure), has drastically reduced its explosion resistance in terms of the 2-nd and the 3-rd criteria.

Figure 4 presents characteristic time histories of strains for samples from different test batches under different exposure levelsβ: 0.44 % (Q=8 g, δ=7.32 mm); 1.04 % (Q=20 g, δ=8.75 mm); 1.64 % (Q=15 g, δ=3.66 mm); 3.27 % (Q=60 g, δ=7.36 mm). For each test (β value) the figure gives readings of 2-3 strain gauges installed at different points over a circle of r=125 mm (see figure 2).

Regarding the strain measurements the following should be noted. For the explosion power range under consideration it is seen that in spite of damage accumulation (failure of binder, fiber ruptures), in general, the test samples do not practically show residual strains at the measurement points or residual deflections, which makes it possible to further model GRP as a linear elastic. The transitory process of sample straining lasts 2-4 ms. This duration depends on both frequency parameters of the samples and the explosion loading level. The time of transitory process decreases as frequency and explosion loading level grows. The latter circumstance is related to the increasing effect of chain radial strains. The maximum radial strains at the measurement points are ~2-2.5 times higher than the circumferential ones. At the explosion loading level initiating the fiber rupture in GRP samples (β=3.27 %), the maximum radial strains reach 2.8 %. At the center of the sample, where the strains are the greatest, this value is even higher and exceeds the limit relative elongation of the material under consideration (2.6-2.8 %, see Table 2). The strain measurement data was used to validate the computer-based analytical models.

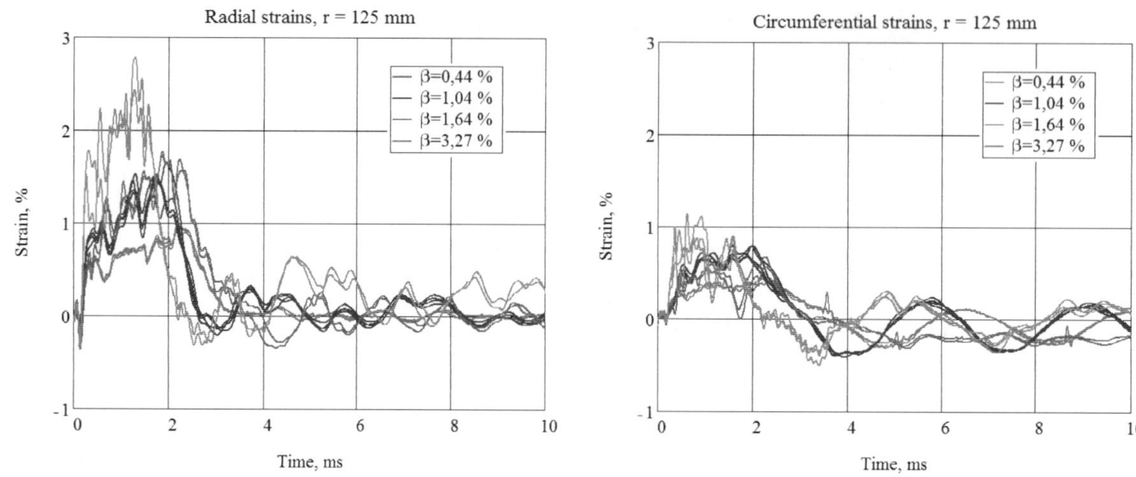

Figure 4. Strain versus time at different explosive loading levels β

Figure 5 shows a typical time history of pressure in the explosion chamber during the tests. Analysis of the pressure records showed that parameters of dynamic sample loading were not affected by any wave reflections from the test chamber walls. The experimental pressure curve was compared versus the calculated parameters of a TNT explosion in the infinite fluid. The calculation was performed using the empirical relationships from [13, 14]:

$$p(t) = p_m \begin{cases} e^{-t/\theta}, & t \le \theta \\ \dfrac{0.368}{t/\theta}, & t > \theta \end{cases},$$

where $p_m = 52.4\left(\dfrac{\sqrt[3]{Q}}{r}\right)^{1.13}$, MPa, $\theta = \dfrac{r_0}{c_0}\sqrt{\ell g\left(\dfrac{r}{r_0}\right) - 0.9}$, s, $r_0 = 0.053\sqrt[3]{Q}$, m, $c_0 = 1500$ m/s – velocity of sound in water, r – distance from explosion center to measurement point, m.

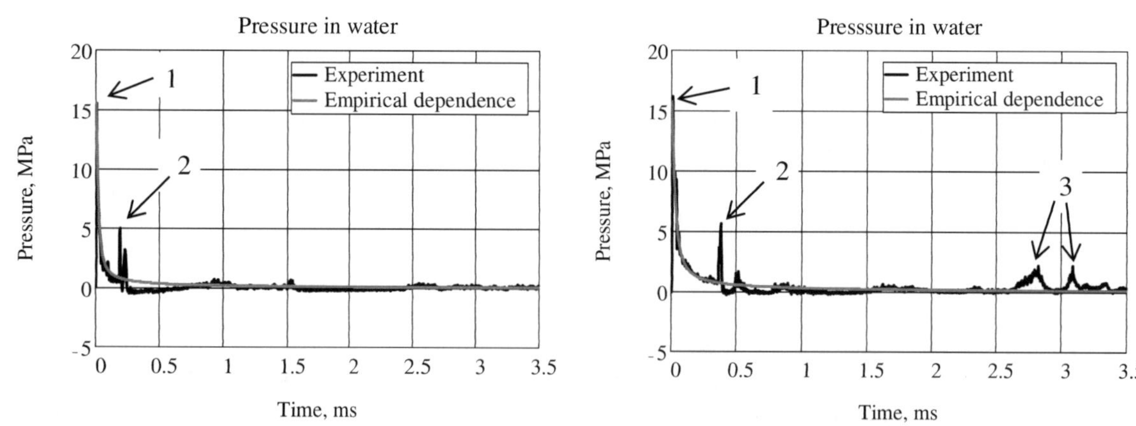

a) $Q = 8$ g, distance to the gauge $r = 0.62$ m *b)* Q = 45 g, distance to the gauge $r = 1.04$ m

Figure 5. Experimental and analytical pressure of explosion: 1 – shockwave front, 2 – wave reflected from the sample, 3 – wave reflected from the chamber walls

The comparison of the test results with the analytical data shows that the charge detonation during the experiments was complete, and the explosion parameters were in a good correlation with the empirical relations for TNT charges, which enabled us to assume Jones-Wilkins-Lee (JWL) equation with empirical TNT parameters for the test explosive charges in our computer-based simulation of the experiment [15].

Computer simulation of the experiment

The purpose of the computer-based simulation was to develop analytical FEM-based models for adequate prediction of strain behavior for composite materials exposed to UNDEX, as well as to apply these models for a more detailed stress-strain analysis of the samples during the tests. The necessity to apply computer-based models in stress analysis is required, among other reasons, due to limited availability of experimental sample strain data: in particular, when the explosion occurs near the sample, strains are practically impossible to record on the front side of the sample, and even on its back side (at the center of the sample).

Computer simulations of UNDEX and its effects on obstacles have been discussed in sufficient detail in many papers. Specifically, various aspects of computer simulations using commercial and in-house software were considered in [16-25], etc.

The processes invoked by a proximity non-contact UNDEX are quite complicated, so in this paper the test conditions were simulated using two FEM-based programs: LS-DYNA and AUTODYN. The formulation of the problem was three-dimensional in both cases. Since the samples were symmetric, only a quarter of the test area was simulated: a segment of fluid (water) with radius of 660-900 mm and height of 810-1050 mm with the cut-out in the upper part running along the contour, its size being equal to overall dimensions of the tubing (see figure 6). The sample was put inside the cut-out, and above the sample, there was air-filled space corresponding to the free volume in the tubing behind the sample. The design diameter of the sample was assumed as equal to the bolt joint diameter, i.e. 500 mm.

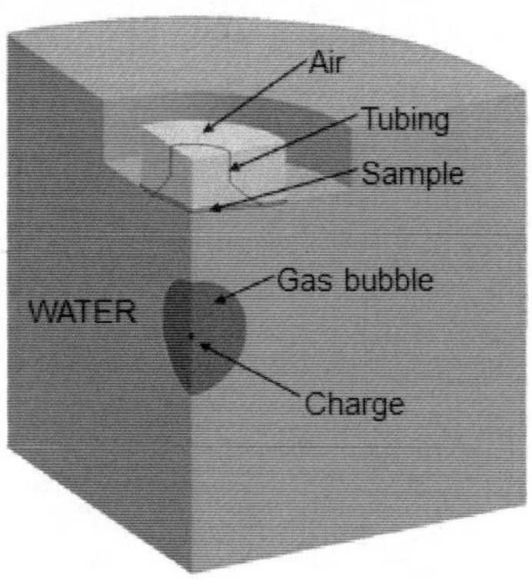

Figure 6. Computational domain in the test simulation

Simulation in both LS-DYNA and AUTODYN was performed with the same boundary conditions and the same types of finite elements. The boundary condition of zero normal velocity was specified for the air-filled space, the contour simulating the tubing was assumed to have solid and zero-normal-velocity boundary conditions, the water space was assumed to have the free-stream boundary. The boundary condition for the sample was specified as "no displacements normal to the sample plane throughout the 50 mm wide ring", which ensured the same test area of the sample as in the actual tests. All the objects – sample, water, air, charge – were simulated by means of SOLID-type finite elements, the Euler mesh was used for water, air and charge, and Lagrange mesh – for the sample.

Mesh parameters (size, aspect ratio, density) were selected so as to have little effect upon the calculation results. In particular, characteristic size of the Euler mesh for water and air near the sample was ~4-5 mm, the size of the Lagrange mesh in the plane of the sample was 3.4 mm, the characteristic number of elements through the thickness of the sample was 6.

Equation of condition for the water was polynomial:

$$p = A_1\mu + A_2\mu^2 + A_3\mu^3 + (B_0 + B_1\mu)\rho_0 e \qquad \text{at } \mu > 0 \text{ (compression)},$$

$$p = T_1\mu + T_2\mu^2 + B_0\rho_0 e \qquad \text{at } \mu < 0 \text{ (tension)},$$

where $\mu = \rho/\rho_0 - 1$, ρ, ρ_0=1000 kg/m^3 – current and initial densities, respectively, A_1, A_2, A_3, B_0, B_1, T_1, T_2 – empirical constants, e – internal energy per unit mass. The following constant values were assumed A_1=2.2 MPa, A_2=9.54 MPa, A_3=14.57 MPa, B_0=B_1=0.28, T_1=A_1, T_2=0. The initial internal energy with reference to the atmospheric pressure of 101.3 kPa was assumed as e_0=361.8 J/kg. It was assumed that the pressure in fluid cannot be negative. This assumption is the simplest way to consider cavitation phenomena in fluids due to explosion processes.

Air was modelled by the equation of ideal gas state $p = (\gamma - 1)\rho e$, where the adiabatic exponent is γ=1.4. It was assumed that ρ_0=1.225 kg/m^3.

The explosion products were simulated using JWL equation

$$p = A\left(1 - \frac{\eta\omega}{R_1}\right)e^{-\frac{R_1}{\eta}} + B\left(1 - \frac{\eta\omega}{R_2}\right)e^{-\frac{R_2}{\eta}} + \omega\rho e,$$

where $\eta = \rho/\rho_0$, ρ_0=1630 kg/m^3, A, B, R_1, R_2, ω – empirical factors. Empirical factors and Chapman-Jouguet parameters for TNT were assumed as [15]: A=371.2 GPA, B=3.231 GPA, R_1=4.15, R_2=0.95, ω=0.3, p_{CJ}=21.0 GPA, D_{CJ}=6930 m/s – detonation velocity, $\bar{e}_0$ = 7000 MJ/m^3 – initial internal energy per unit volume ($e_0 = \bar{e}_0/\rho_0$ = 4.29 MJ/kg).

The underwater explosion effect upon the sample was simulated in two stages. The first stage was the investigation of an explosion in the free infinite fluid in uni-dimensional (AUTODYN) or two-dimensional (LS-DYNA) formulation. In this formulation, the calculation was performed up to the moment corresponding to the shock wave approach to the sample. Then, this solution was exported into a 3D model including the sample, and the second stage of calculation followed.

GRP was treated as a quasi-isotropic material with linear elastic properties. For valid comparison with the test results the sample characteristics (thickness, density, modulus of elasticity, etc.) were assumed to be the same as in the test under consideration

Computer models were validated using the strain time histories recorded during 9 tests on batches 1 to 4. The mass of charge in the experiments was Q = 8, 10, 15, 45 and 60 g. Figure 7 compares analytical and experimental time histories of radial and circumferential strains for a sample of batch 1 (δ=8.75 mm) at explosion of charge Q=10 g.

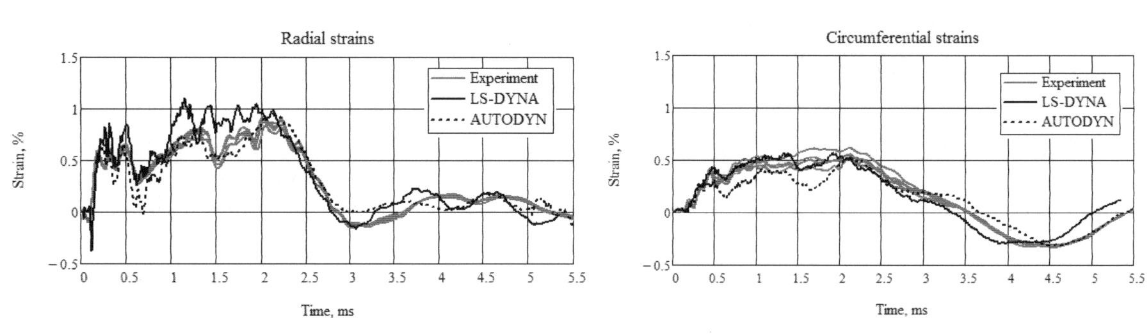

a) radial strains (RC=0.14 – for LS-DYNA; RC=0.18 – for AUTODYN)

b) circumferential strains (RC=0.10 – for LS-DYNA; RC=0.22 – for AUTODYN)

Figure 7. Calculation vs experiment: Q=10 g, δ=8.75 mm

Quantitative assessment of the convergence between the calculation results with the test data was performed as per the approach suggested and justified in [26]. As per this approach, integrated coefficient of convergence between two unsteady processes (e.g. experimental and analytical) is determined as follows:

$$RC = \sqrt{\frac{\pi}{4}\left(RM^2 + RP^2\right)},$$

where $RM = sign(m)\,\ell g\left(1+|m|\right)$ and $RP = \dfrac{\cos^{-1}(p)}{\pi}$ – divergence by amplitude and by phase, respectively, $m = \dfrac{A-B}{\sqrt{AB}}$, $p = \dfrac{C}{\sqrt{AB}}$, $A = \sum_i f_1(i)^2$, $B = \sum_i f_2(i)^2$, $C = \sum_i f_1(i) f_2(i)$, $f_1(i)$ and $f_2(i)$ – amplitudes of the experimental and simulated process, respectively, as determined for the same time instant i. Ref. [27] suggests the following gradation for the integrated RC coefficient: if $RC \le 0.15$, the convergence is deemed excellent, if $0.15 < RC \le 0.28$, the convergence is deemed acceptable, and if $RC > 0.28$, the convergence is deemed poor.

Figure 8 gives calculated *RC* coefficient for radial and circumferential strains in 9 experiments with GRP samples (triangles and circles denote the calculation results as per LS-DYNA and AUTODYN respectively) [11]. As it is seen from this figure as well as from figure 7 calculation results have rather a good correlation with the experiment. Only in 4 cases for LS-DYNA and in 2 cases for AUTODYN the coefficient was $RC > 0.28$. Some deviation between the calculation and the experiment, as well as between the calculations in LS-DYNA and in AUTODYN can be due to different accuracy in evaluation of explosion parameters at the initial stage between LS-DYNA and AUTODYN, errors in modelling boundary conditions for samples, evaluation of actual sample thickness, discrepancy in coordinates of strain and locations of strain gauges.

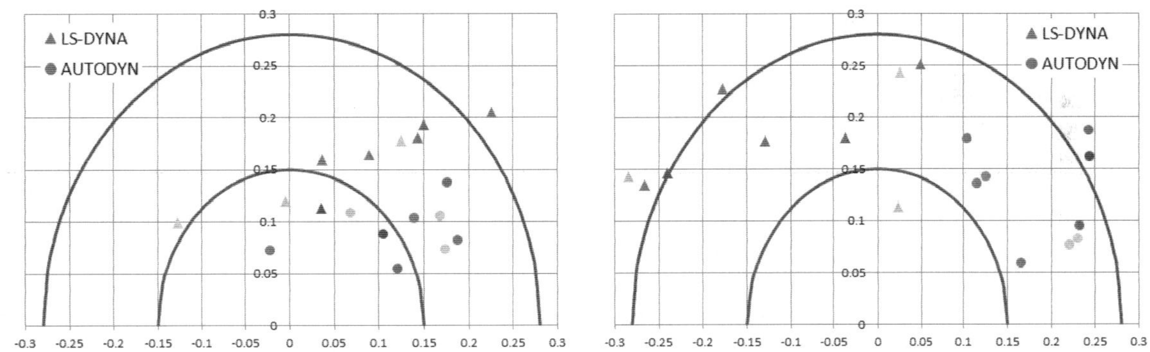

a) radial strains b) circumferential strains

Figure 8. Integrated coefficient of convergence between calculations and experiment

The developed computer models were used for more accurate stress analysis of the samples during the tests. Figures 9 and 10 illustrate estimates of sample deflection behavior in the process of deformation. Three phases of the process can be identified. First (at the initial instant at t=0.1 ms, see figure 10b) the sample is mainly deformed around supports by bending and shear. The middle part is practically intact and moves as a solid body. In the second phase, deformation waves (shear, bending and tension) propagating from the supported contour reach the central part of test sample. In this case the sample assumes a sine-like shape along its radius (figure 10). Further, in Phase 3, the sample continues to deform mainly retaining the same shape of deflection. The local peaks of deflections in the center of test sample (figure 9) are caused by appearance and disappearance of cavitation areas on the sample. The above phases of deformation process are characteristic for the conditions when the sample surface is subjected to a distributed dynamic load of shock wave type [28].

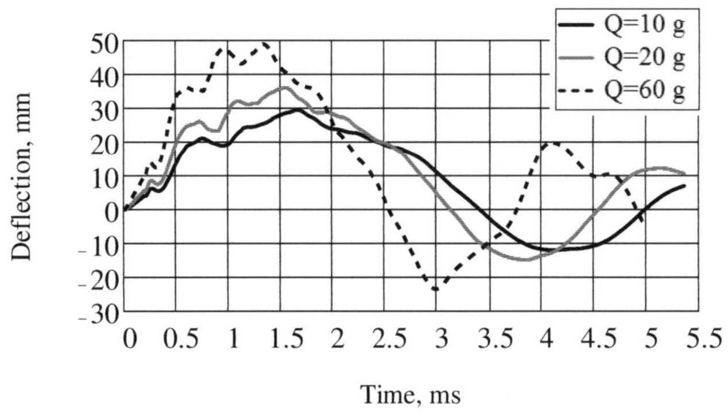

Figure 9. Deflection of sample center in function of time, δ=8.75 mm

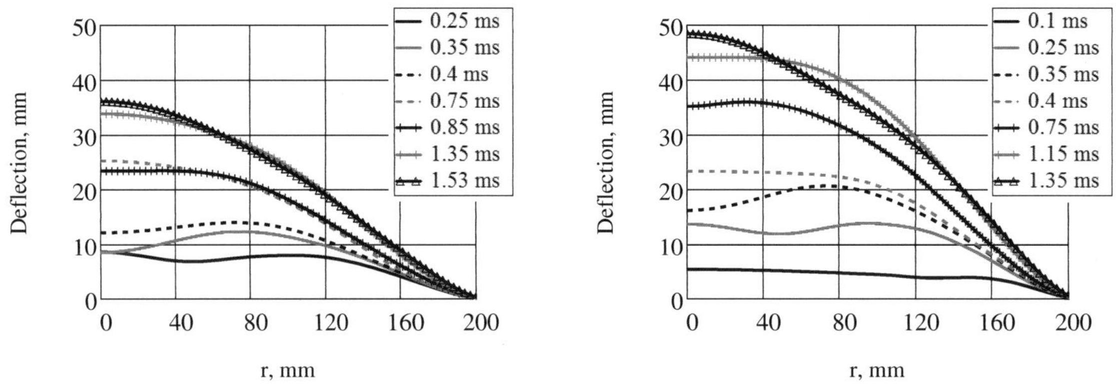

a) Q=20 g b) Q=60 g

Figure 10. Shape of sample deflection (along radius) at different instants in time, δ=8.75 mm

An important consideration in the limit state estimations is the level of sample strain. The chain and bending strains (figure 11), as well as strain intensities (figure 12) in the sample center and at the support were considered. The analysis of calculation results shows that even the samples of relatively large thickness (batch 1, δ =7.91-8.75 mm) feature a high level of chain strains. The maximum value of chain strains is practically equal to bending strains, and is approximately 2 % at 60 g charge explosion (tentatively initiating rupture of individual fibers). In this case the sample deflection is about 50 mm. Bending strains are prevalent in the supported area, exceeding 5 to 7 times the level of chain strains. The maximum level of total radial (bending + chain) strain at the support is about 1.5 times higher than at the sample center.

In accordance with the test data, samples are first damaged in the center (rupture of fibers) in spite of higher radial strains at the support. However, if we compare the strain levels by von Mises criterion these strains are approximately 1.5-2 times higher at the center as compared to the support. At the initiation of fiber rupture von Mises yield reach 5 % in the center of samples, which is twice the ultimate relative elongation of static samples. It can be assumed that the von Mises yield stress can be used to evaluate the explosion resistance (based on fiber rupture criterion) for the polymer composites manufactured by vacuum infusion process and featuring quasi-isotropic properties provided by their reinforcement structure.

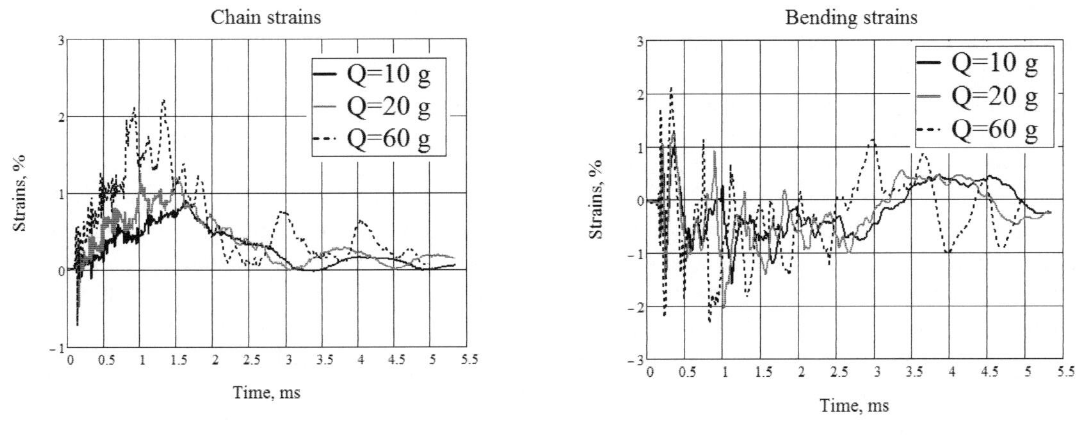

a) strains in the center

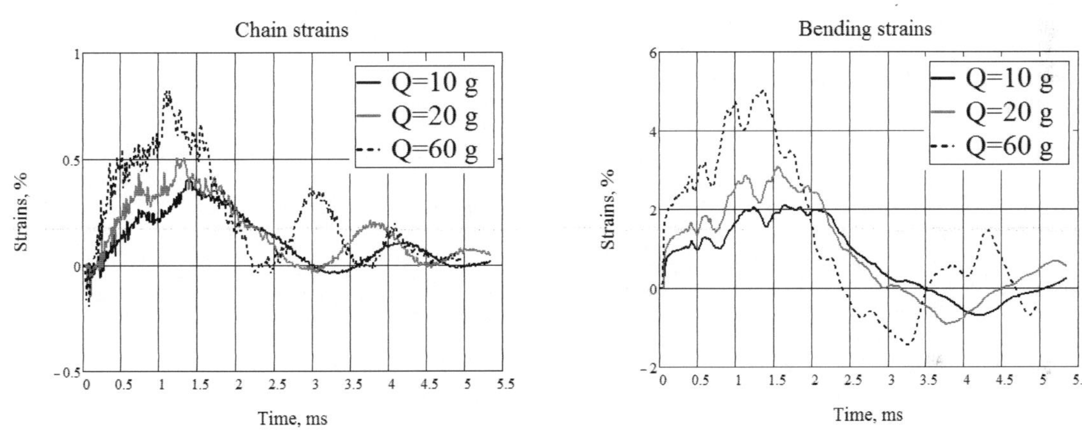

b) strains at the support

Figure 11. Chain and bending strains: calculation results, δ=8.75 mm

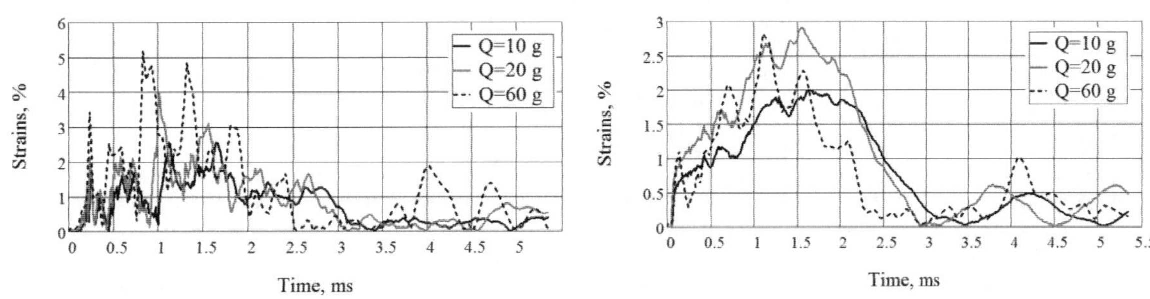

a) in the center b) at the support

Figure 12. von Mises yield strains: calculation results, δ = 8.75 mm

Conclusion

The paper presents the results of experimental and computational studies conducted to investigate underwater explosion resistance of GRP samples made of quadraxial glass fiber and vinyl ester binder. The samples were tested in the explosion test chamber of Krylov State Research Centre. The sample material damage was assessed versus three indicative effects (criteria): binder failure, fiber rupture, through-thickness penetration (hole). The relative explosive charge mass (charge mass per unit effective area of sample related to sample surface mass) is taken as a measure of explosion effects causing a certain type of material damage. It is confirmed experimentally that the specific explosion resistance of the test samples up to through-thickness penetration is largely defined by the energy level initiating the fiber rupture process. Computer simulation models were developed and validated to enable more detail analysis of stress-strain states of tested samples. In particular, the computer simulations indicate that the von Mises yield criterion can be used to assess the explosion resistance (initiation of fiber rupture) for samples under study. It should be noted that von Mises strain exceeds approximately two times the ultimate relative elongation of static samples.

The above-presented experiment in modelling UNDEX effects should be useful for the analysis of stress-strain states and failure processes of samples and structures made of various GRP materials.

References

[1] Mouritz A 1995 *Composites* **26** (1) 3–9
[2] Batra R and Hassan N 2007 *Composites* B **38** 448–68
[3] LeBlanc J and Shukla A 2010 *Composite Structures* **92** (10) 2421-30
[4] Avachat S and Zhou M 2011 *Proc 18th Int Conf on Composite Materials* (21-26 August 2011 S. Korea) W26-6-AF2057
[5] LeBlanc J and Shukla A 2011 *Int J Impact Eng* **38** 796-803
[6] LeBlanc J and Shukla A 2011 *Composite Structures* **93** 3072–81
[7] Dear J Arora H and Hooper P 2011 *Proc 18th Int Conf on Composite Materials* (21-26 August 2011 S. Korea) W26-3-AF0730
[8] Schiffer A and Tagarielli V 2014 *Int J Impact Eng* **70** 1-13
[9] Schiffer A and Tagarielli V 2015 *J of Fluids and Structures* **52** 130-144
[10] Dulnev A and Nekliudova E 2017 *Tomsk State University Journal of Mathematics and Mechanics* **47** 51-62 (*in Russian*)
[11] Dulnev A, Nekliudova E and Weiss I 2017 *Transaction of the Krylov State Research Center* **2** (380) 54-64 (*in Russian*)
[12] 2001 Destruction of objects of different scales under explosion ed A G Ivanov (Russian Federal Nuclear Center-VNIIEF Sarov) p 482 (*In Russian*)
[13] Cole R 1948 *Underwater explosions* (Princeton: Princeton University Press) p 435
[14] Zamyshlyaev B and Yakovlev Yu 1967 *Dynamic loads during the underwater explosion* (Leningrad: Sudostroyeniye Publishing House) p 387 (*in Russian*)
[15] 2002 *Physics of the explosion* vol 1 ed L Orlenko (Moscow: Fizmatlit Publishing House) p 832 (*in Russian*)
[16] Kormilitsin Yu N, Melnikov S Yu and Tomashevsky V T 2006 *Underwater explosion and its interaction with media and barriers* (Saint-Petersburg: Nauka Publishing House) p 243 (*in Russian*)
[17] Dombrosky S E 2003 *Int Conf Zababakhin Scientific Talks* (8-12 September 2003 Snezhinsk Russia)
[18] Dombrosky S E and Palmin S A 2003 *Int Conf Zababakhin Scientific Talks* (8-12 September 2003 Snezhinsk Russia)
[19] Yanilkin Yu V *et al* 2014 *Int Conf Zababakhin Scientific Talks* (2-6 June 2014 Snezhinsk Russia)
[20] Hammond L and Grzebieta R 2000 *Shock and Vibration* **7** 333-41
[21] Shin Y S 2004 *Computers and Struc*tures **82** 2211-19
[22] Sprague M A and Geers T L 2006 *Comput Methods Appl Mech Eng* **195** 2149-67
[23] Qiankun J and Gangyi D 2011 *Int J Impact Eng* **38** 558-66
[24] Glazyrin V P, Orlov Yu N and Orlova Yu N 2009 *Izvestia vyshikh uchebnykh zavedeniy* **52** issue 7/2 77-80 (*In Russian*)
[25] Glazyrin V P, Orlov Yu N and Orlova Yu N 2009 *Izvestia vyshikh uchebnykh zavedeniy* **52** issue 7/2 74-77 (*In Russian*)
[26] Russell D 1997 *Proc 68th Shock and Vibration Symp* (3-6 November 1997 Hunt Valley Maryland)
[27] Russell D 1998 *Proc. 69th Shock and Vibration Symp* (12-19 October 1998 Minneapolis/St. Paul Minnesota)
[28] Slepyan L I 1972 *Non-stationary elastic waves* (Leningrad: Sudostroyeniye Publishing House) p 374 (*in Russian*)

The study of the impact of ice conditions on the possibility of the submarine vessels surfacing in the ice cover

V L Zemlyak[1], V M Kozin[1,2], N O Baurin[1], K I Ipatov[1], M V Kandelya[1]

[1]Sholom-Aleichem Priamursky State University, Birobidzhan, Russia
[2]Institute of Machining and Metallurgy, FEB RAS, Komsomolsk-on-Amur, Russia

E-mail: vellkom@list.ru

Abstract. Traditionally submarine vessels emerging from under ice cover performing by static loading of ice from bellow through the creation of positive buoyancy by main ballast tanks. However thickness of ice (approximately 1 meter) from under which modern submarine vessel can emerge essentially limits the use of traditional method, particularly during submarine vessels motion in the severe ice conditions of Arctic region. For breaking the ice cover of greater thickness can be used flexural gravity waves caused by the submarine vessel motion with certain critical speed near the bottom ice. It is also known that in the coastal areas water depth is often less than 100 meters, and the presence of projections on the bottom surface may effect on the wave propagation pattern. This paper presents experimental study of influence of bottom contour on the deflection and the length of the flexural gravity waves from the movement of submarine. Ice cover failure pattern determined. Assessment of ice-breaking capacity of flexural-gravity waves with using the criterion of ice failure is performed.

1. Introduction

Traditionally submarine vessels emerging from under ice cover performing by static loading of ice from bellow through the creation of positive buoyancy by main ballast tanks. However thickness of ice (approximately 1 meter) from under which modern submarine vessel can emerge essentially limits the use of traditional method, particularly during submarine vessels motion in the severe ice conditions of Arctic region. For breaking the ice cover of greater thickness can be used flexural gravity waves (FGW) caused by the submarine vessel motion with certain critical speed near the bottom ice. Among the first theoretical papers on the subject is a paper by Kheisin [1], who studied the plane steady-state problem of the motion of a vortex under a layer of broken ice. Bukatov and Zharkov [2] and Kozin and Pogorelova [3] studied the steady motion of a point source of mass under a floating elastic plate, analyzed the effect of the velocity of motion, the depth of the source, and the thickness of the plate on its deflections. The possibility of breaking natural ice cover by flexural gravity waves (FGW) caused by motion of a model submarine vessel with the relative hull extension $L_m/B_m = 8$ (where L_m is the length of the vessel, B_m is the width of the vessel) and the full-scale vessel displacement D_n=6000 t after converting to the natural one was shown in the work Kozin et al [4]. The destruction of ice by dynamic loads is considered in the works [5, 6]. In its turn, the presence of protruding parts, as well as the form of the cross section of submarines of various projects of can have a significant influence on the parameters of flexural gravity waves [7]. It is also known that in the coastal areas water depth is often less than 100 meters, and the presence of projections and ledges on the bottom surface may effect on the wave propagation pattern.

2. Equipment and technique for conducting experiments

The aim of the study was experimental determination of the impact of the bottom contour on the model ice failure behavior, length and deflection of flexural-gravity waves generated by the motion of the immersed body.

Experimental studies of submarine vessel model motion under the ice sheet at a limited water depth were carried out in the natural ice basin with the following size 10×3×1 m laboratory of ice technology "Amur State University after Solom-Aleikhem" (Birobidzhan, Russia) [8]. Double bottom was modeled by using three sections mounted on the movable modules allowing to set required water depth by use of computer control. The block diagram of the basin is shown in figure 1. The chart of projections is shown in figure 2.

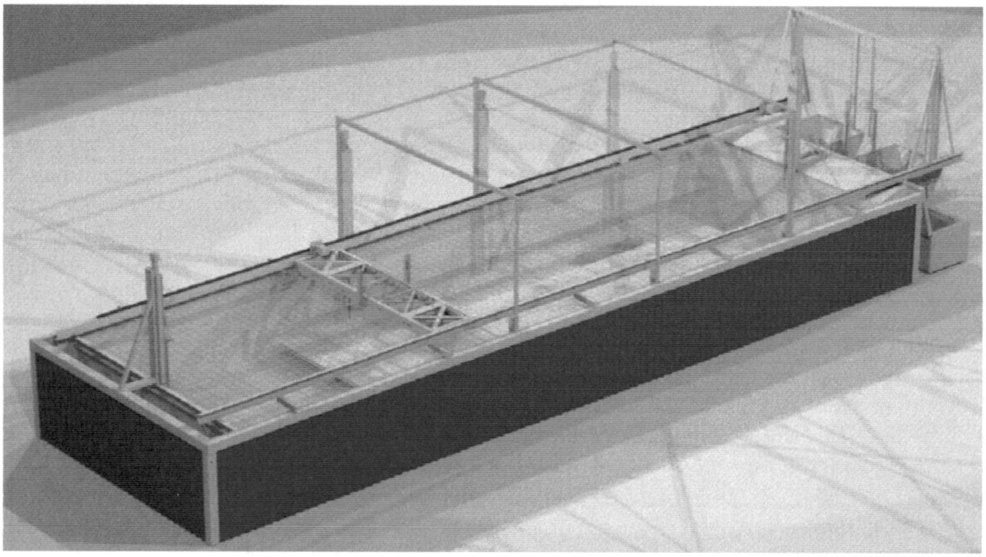

Figure 1. General view of the ice basin

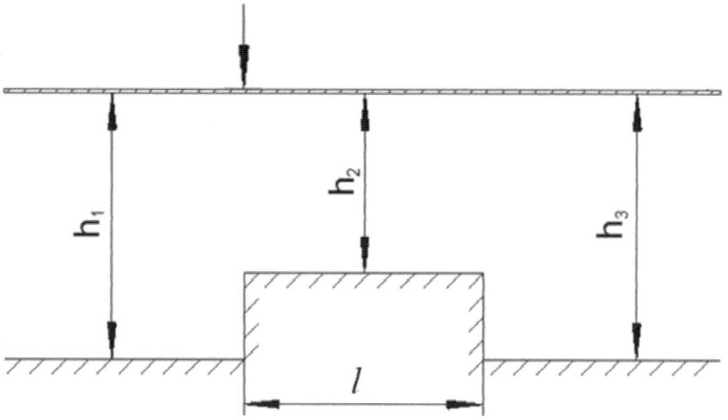

Figure 2. The chart of simulated projections of the bottom (the pointer indicates the location of vertical displacement sensor) (h_1=h_3=0.4 m, h_2=0.3÷0.37 m, l=2 m)

Due to the lack of refrigerating machines, experimental studies were carried out only in the winter period (December-January). The model of the ice cover was made in the ice basin by freezing natural freshwater without any dopant up to the set thickness (0.003 m) in natural conditions. The room where the basin is located was not heated for the entire period of the experiment. The experiments were carried out only at night time from 9.00 p.m. to 5.00 a.m. At night there were negligible fluctuations in the room temperature, within the difference of 1°C to 2°C, although the room temperature varied from -10°C to -15°C, depending on the outside temperature, varying from -25°C to -30°C. Relatively uniform ice sheets were thus obtained in 2-2.5 hours at night, except for the patches of thaw on the ice. Effects of snow and wind from the outside were minimized at the ice basin. Due to a layer of insulating material installed on the sides and under the bottom of the basin and compliance with constant temperature mode, the thickness of the ice at the walls of the basin and in the middle of the ice field did not exceed 0.001 m. After carrying out of each model test, broken ice pieces were removed and the depth of the model was readjusted, and then ice sheet was formed.

To conduct the experiments necessary equipment and measuring system to register vibrations of model ice was prepared. By towing the model vessel its velocity and profiles of FGW were determined. To record the vibrations of the model ice we used infra-red non-contact vertical travel sensors LAS-Z by Way Con (Germany). The profile of the waves generated by the moving model was recorded with the help of the program Test viewer 2.34. The towing system allowed carrying out model tests at velocities of up to u_m=2.4 m/s under steady motion conditions.

Submarine vessel model scale equal to λ_l=1:120 was made as drop-shaped body of rotation, length-diameter ratio L_m/B_m=8.4 (where L_m=1.15 m). The length of parallel middle body was L_{pm}=0.7 m. Full underwater displacement after expansion was equal to D_n=24000 t. The parameters of the model were chosen considering previously executed experimental studies in view of modeled ice conditions these parameters are optimal.

To give it a zero buoyancy and positive stability when moving under model ice the model vessel was fitted with solid ballast. To create a technically smooth surface the body of the model vessel was puttied and painted. Harnessing of the model in the towing system was made through carabiners connecting the tow cable with attachment points located in the extremities of the body.

Simulation scale equal was selected basing on the size of the ice pool, with the length of the channel providing access to the steady motion of the model SV. The width was sufficient to eliminate the influence of the waves reflected from the channel sides, on the principle wave system [9]. Modeling of sea ice cover can be executed by using various ice models and for each of them there are corresponding conditions of similarity. The usual modeling is performed with partial satisfaction of conditions of similarity [10]:

$$\lambda_E = \lambda_w = \lambda_h = \lambda_l \qquad (1)$$

Where λ_E is the model scale for Young's modulus; λ_w is the model scale for deflections; λ_h is model scale for ice thickness.

In this case, the requirements for similarity conditions with respect to Poisson's ratio μ and density ρ of the model ice are satisfied.

The model vessel must be geometrically similar to the real one and their displacements must be in proportion to the cubed modulus of the geometrical scale-ratio:

$$L_n/L_m = \lambda_l, \qquad D_n/D_n = \lambda_l^3 \qquad (2)$$

where L_n is the length of the full-scale vessel; D_m is the model vessel displacement.

The model motion velocity u_m is determined by the condition of similarity:

$$u_n/u_m = \lambda_l^{1/2} \qquad (3)$$

where u_n is the full-scale vessel velocity.

The parameters of the model FGW are converted to natural ones in accordance with the modulus of geometrical similarity:

$$\lambda_n/\lambda_m = w_n/w_m = \lambda_l \tag{4}$$

where λ_n is the length of the full-scale FGW; λ_m is the length of the model FGW; w_n is the full-scale FGW deflections; w_m is the model FGW deflections.

When natural ice cover is used as model one, the thickness of the modeling ice will be calculated when converted to natural one in accordance with the following relations:

$$h_n = h_m \lambda_l^{4/3} \left([\sigma_u]_n \big/ [\sigma_u]_m \right)^{-\frac{1}{3}} \tag{5}$$

where h_n is the natural ice thickness; h_m is the modeled ice thickness; $[\sigma_u]_n$ is the natural flexural stresses; $[\sigma_u]_m$ is the modeled flexural stresses.

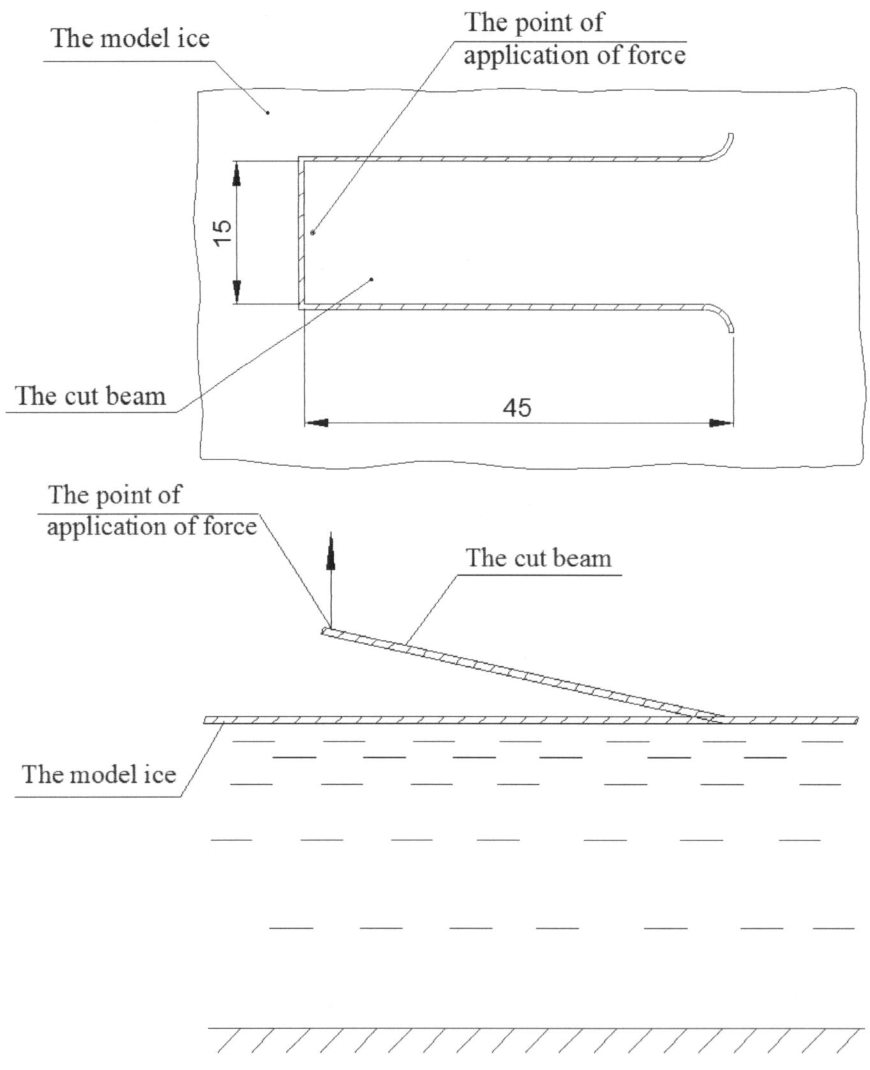

Figure 3. Testing beams

Determination of flexural strength of the model ice was experimentally performed by testing beams and for that purpose the model ice cover was prepared by building up ice of the required thickness h_m=0.003 m [11]. The beams had a rectangular shape with the parameter of $l \times b$=0.15×0.45 m which were prepared in the model ice by cutting out according to the stamp by the rotating Dremel tool. The force required for beam destroying was determined by using the electronic dynamometer Mark - 10 (USA) (figure 3). The average value of the flexural strength for natural ice was the value of $[\sigma_u]_n$=0.7 MPa. The thickness of the modeling ice cover after conversion to natural one was h_n=2.3 m.

It is known that at modeling sea ice by using fresh-water model ice, modulus of elasticity of the natural ice should be less then modulus of elasticity of the model ice $E_n > E_{m,}$ however the ratio E/σ_u for sea ice is practically the same as that for fresh-water ice. To meet this condition at modeling ice is very important [12]. To enhance the visibility of fractures of ice, the model ice surface was covered with a thin layer of snow, soon after the model field formation.

The coefficient α was used as a criterion for breaking model ice, that is it is equal to the maximum value of the tangent of the surface slope of the ice plate. The relationship between failure of ice cover and the coefficient α was already empirically determined in Kozin's experimental work [4]. This series of model experiments on the failure of natural ice cover of various thicknesses were conducted using models of submarine vessels and hovercrafts. It was stated, that for the load motion over the ice cover as well as for motion of a submarine vessel when the maximum value of the slope of the ice surface is greater than the value of 0.04, it leads to complete ice failure and cracks opening. The value of α in experimental studies was defined by the formula:

$$\alpha = 2\pi w_m / \lambda_m, \tag{6}$$

We should note that as far as the value of the criteria was obtained for the complete failure of freshwater ice, its usage for assessment of sea ice failure is not entirely correct. However, if we take into account that ice strength and fracture toughness of sea ice is lower than that of freshwater, the value of α for sea ice would be a bit lower.

Due to the complexity of the study, the duration of the preparation of the model ice field and limited cold period (December-January), aiming to identify optional parameters of the experiment such as model displacement velocity, relative deepening and water depth a series of preliminary test model runs was executed. According to the performed study the most significant results to be expected at velocity of the model vessel equal to u_m=1.25÷2.18 m/s, model deepening equal to h_m=0.16-÷0.32 m and water depth equal to H_m=0.3-0.45 m, which after conversion to natural was: u_n=13.7÷23.9 m/s; h_n=19.2÷38.4 m; H_n=36÷54 m.

3. Results of the model experiments

It is known that the parameters of FGW caused by moving of the load on the ice cover significantly depend on the depth of the basin [13]. The deflections of the ice cover in shallow water is always greater than in deep water, at that the length of FGW increases, propagation velocity decreases and FGW steepness ratio changes. The results of experiments are shown in figures 4-6.

The analysis of the experimental data revealed a significant effect of the water depth on the parameters of the FGW caused by the load motion. Comparing obtained data with the results of experiments for ice sheet at an infinite depth of the basin figure 7, it may be concluded that ice deflections increased.

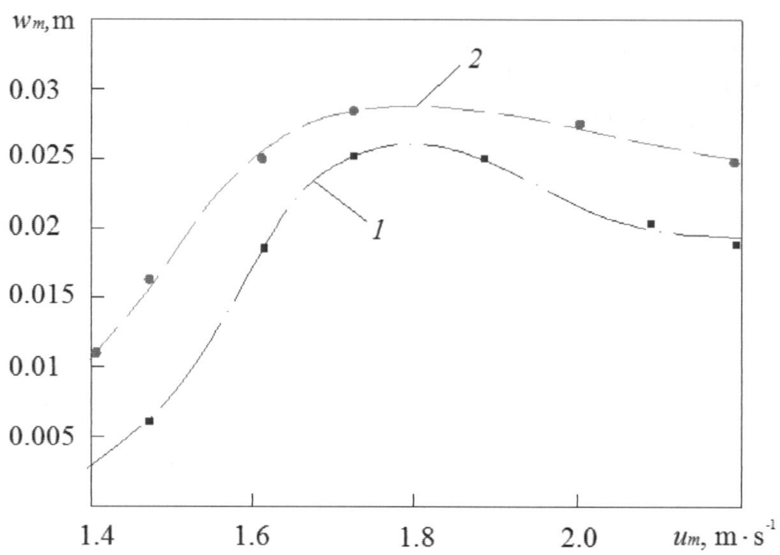

Figure 4. The relationship of the maximum deflections at different SV model velocity at a water depth equal to H_m=0.35 m: 1 depth of the SV h_m=0.16 m; 2 depth of the SV h_m=0.21 m.

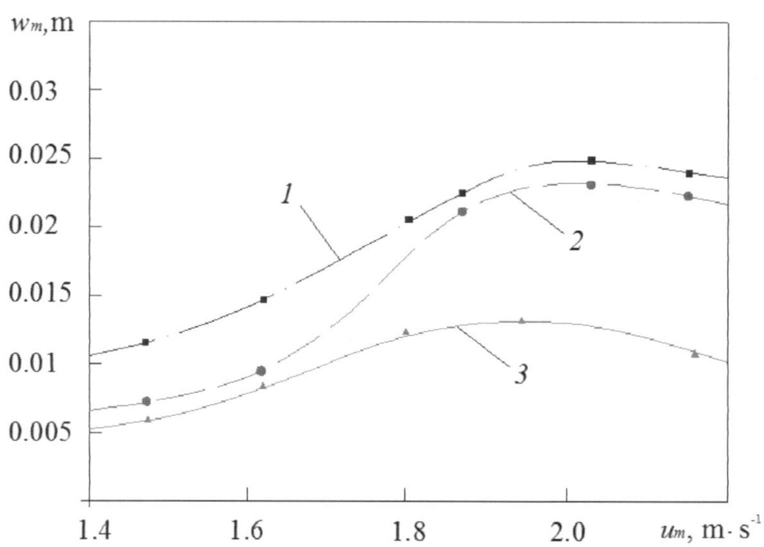

Figure 5. The relationship of the maximum deflections at different SV model velocity at a water depth equal to H_m =0.4 m: 1 depth of the SV h_m=0.16 m; 2 depth of the SV h_m=0.21 m; 3 depth of the SV h_m=0.26 m.

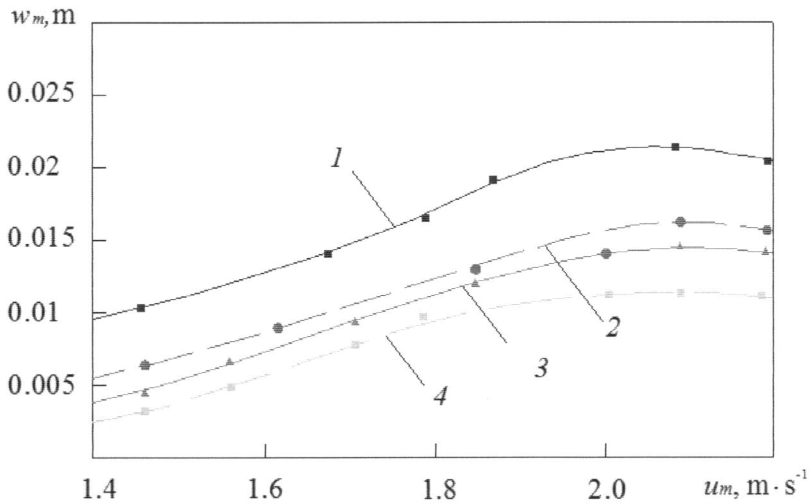

Figure 6. The relationship of the maximum deflections at different SV model velocity at a water depth equal to H_m=0.4 m: 1 depth of the SV h_m=0.16 m; 2 depth of the SV h_m=0.21 m; 3 depth of the SV h_m=0.26 m; 4 depth of the SV h_m=0.31 m.

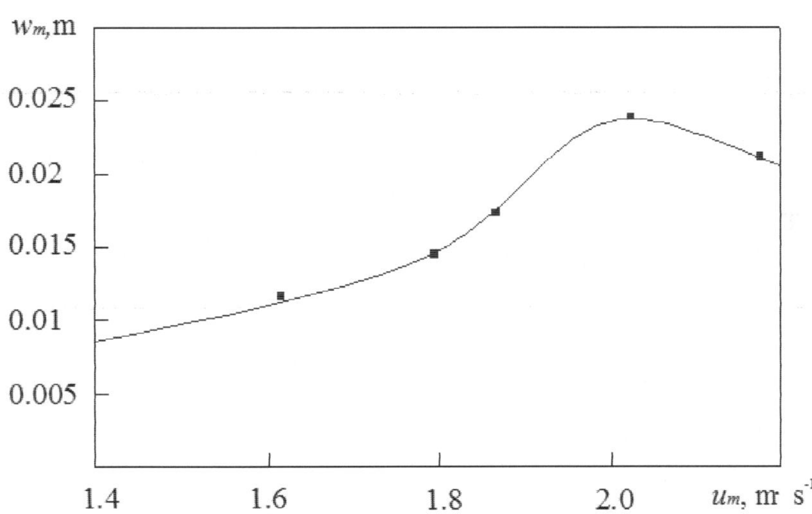

Figure 7. The relationship of the maximum deflections at different SV model velocity at a water depth equal to H_m=1.0 m, h_m=0.16 m.

The relationship between coefficient α and load motion velocity for various depth is shown in the Figs. 8-10. At that minimum velocity of the load motion at which ice failure occurred was much less than in the case of an infinite water depth and was equal to u_m = 1.42 m/s.

IOP Conf. Series: Journal of Physics: Conf. Series **919** (2017) 012004 doi:10.1088/1742-6596/919/1/012004

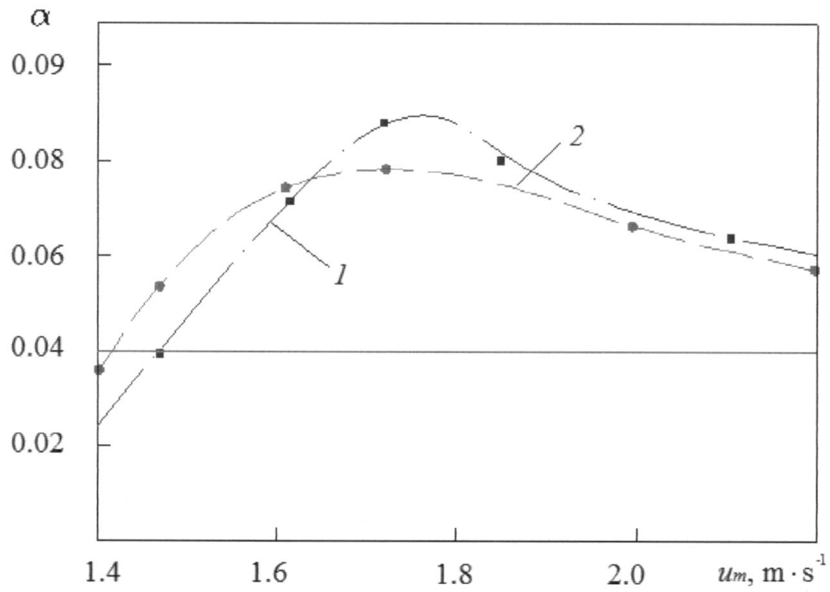

Figure 8. The relationship of the coefficient α at different SV model velocity at a water depth equal to H_m=0.35 m: 1 depth of the SV h_m=0.16 m; 2 depth of the SV h_m=0.21 m.

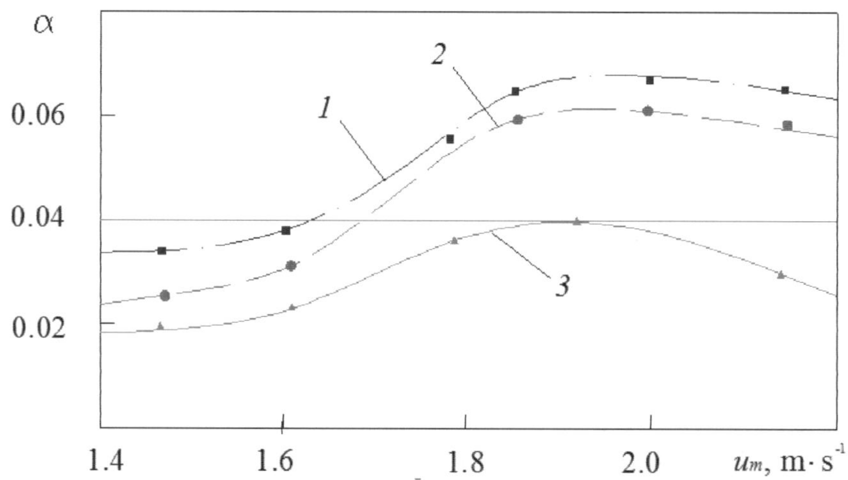

Figure 9. The relationship of the coefficient α at different SV model velocity at a water depth equal to H_m=0.35 m: 1 depth of the SV h_m=0.16 m; 2 depth of the SV h_m=0.21 m; 3 depth of the SV h_m=0.26 m.

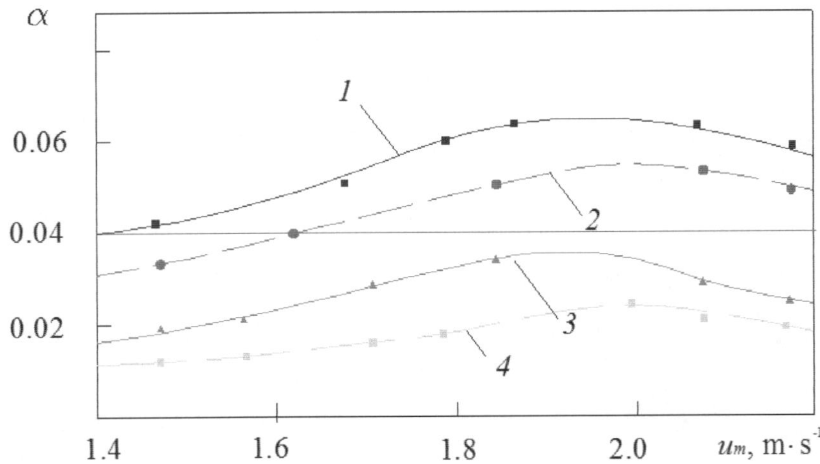

Figure 10. The relationship of the coefficient α at different SV model velocity at a water depth equal to H_m=0.35 m: 1 depth of the SV h_m=0.16 m; 2 depth of the SV h_m=0.21 m; 3 depth of the SV h_m=0.26 m; 4 depth of the SV h_m=0.31 m.

Character of the failure of the model ice is shown in figures 11-13. It is should be noted that most intensive ice failure and ice pieces crushing occured at near-critical velocity. Ice cover completely lost its carrying capacity. When reducing model velocity the area of the fracture was abridged significantly, although the intensity of destruction at low velocity was higher than at a velocity greater than the critical value, which is associated with high curvature of FGW. In the same way when moving in conditions of limited water depth longitudinal crack formed along the line of motion model.

Figure 11. Character of the failure of the model ice by FGW: H_m=0.35 m; h_m=0.16 m; u_m=1.47 m/s

Figure 12. Character of the failure of the model ice by FGW: H_m=0.35 m; h_m=0.16 m; u_m=1.73 m/s

Figure 13. Character of the failure of the model ice by FGW: H_m=0.35 m; h_m=0.16 m; u_m=2.09 m/s

Flexural gravity waves pattern fixed by the sensor on the edge of the projection is shown in figure 14.

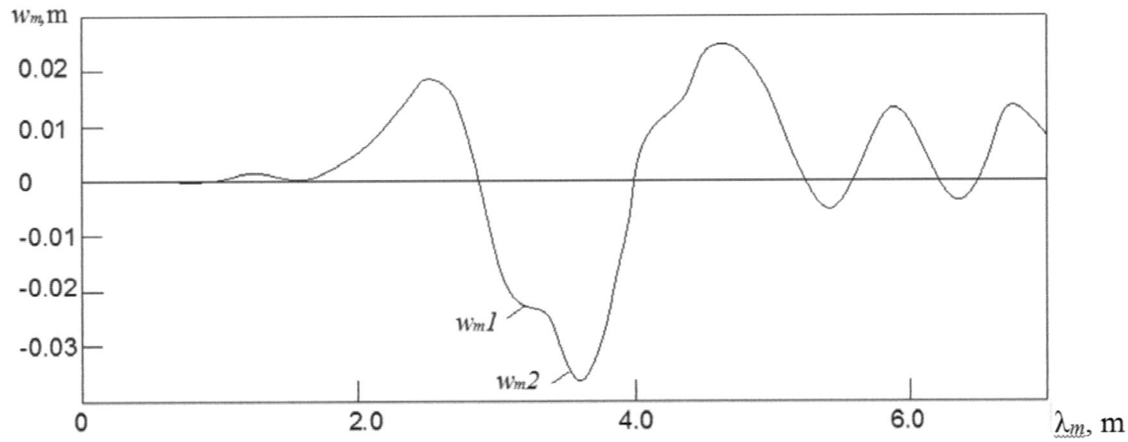

Figure 14. FGW pattern fixed by the sensor on the edge of the projection (the projection №1)

The results of model experiment-s on simulating of projections and ledge of the bottom shown in figures 15-16.

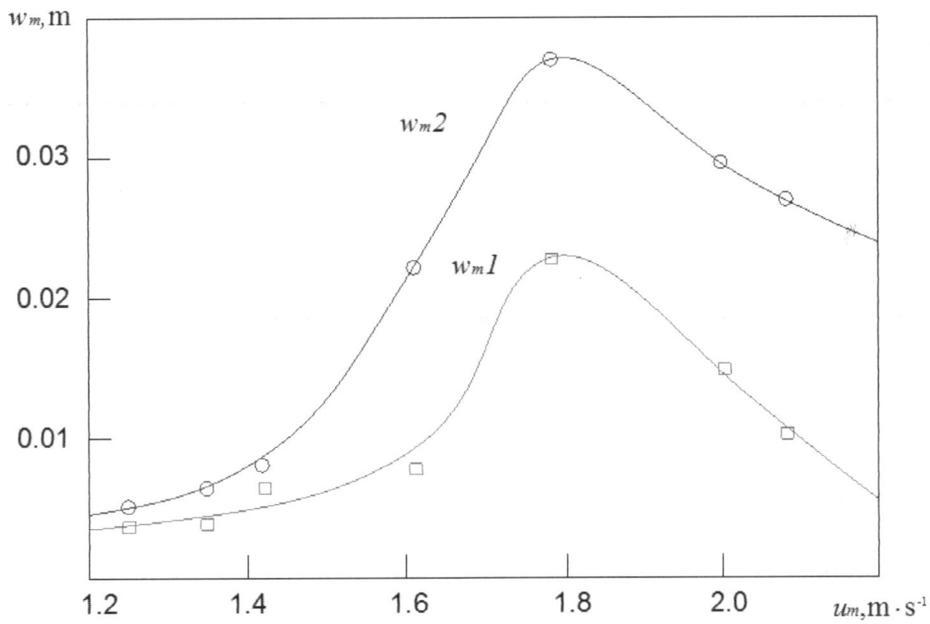

Figure 15. The relationship of the maximum deflections at different SV model velocity of the projection №1

Analysis and comparison of the data demonstrates a significant effect of simulated conditions on the parameters of the modeled FGW. The presence of the projection gives birth to a hydroshock and leads to a sharp increase of curvature of FGW by comparison with a continuous flat bottom. The graphs show that the curvature of the waves increases by 16-33%.

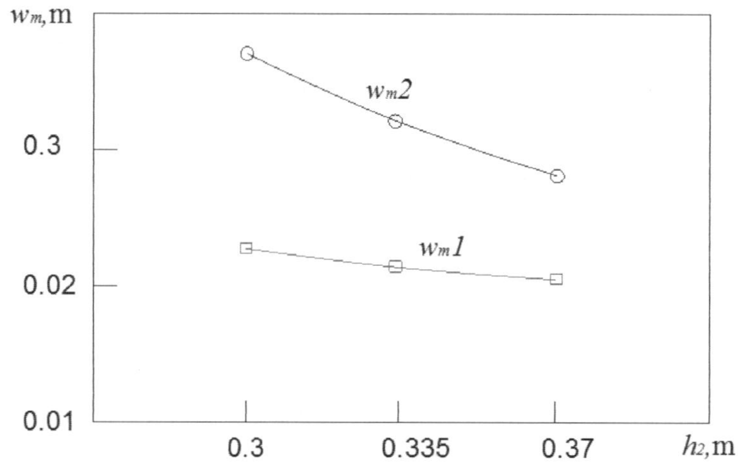

Figure 16. The relationship of the maximum deflections of the height of the projection (h_2)

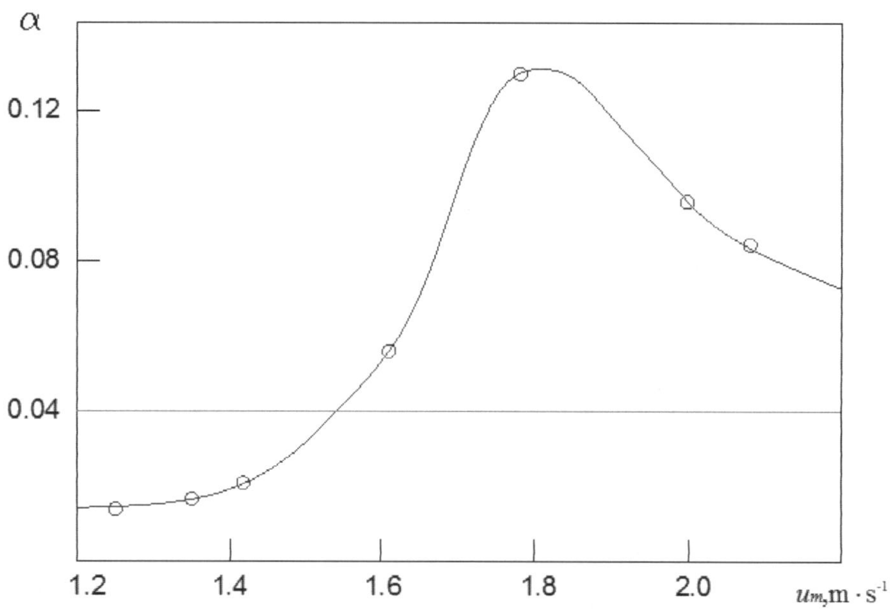

Figure 17. The relationship of the coefficient α at different SV model velocity of the projection №1.

High intensity of the destruction and grinding of ice floes observed when passing by SV model the areas of the simulated contour. The field of simulated ice was losing the bearing capacity completely. The dependence of coefficient α on the movement velocity of the load is shown in figures 17-18.

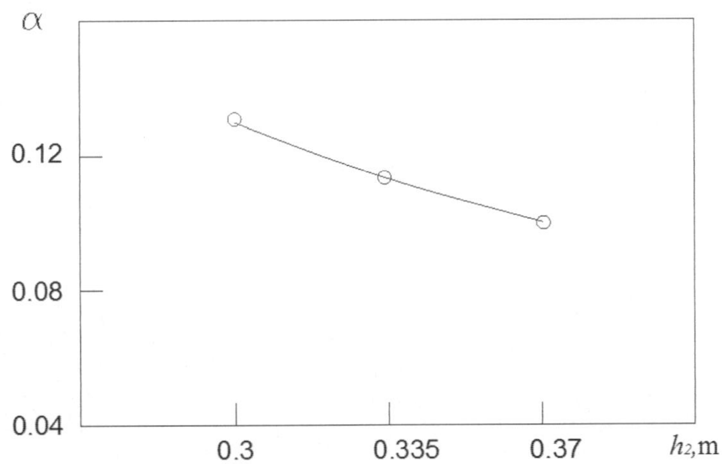

Figure 18. The relationship of the coefficient α at different of the height of the projection (h_2)

The graphs show that the intense destruction of ice occurred in all velocity ranges and types of bottom contour under the study. Moreover, the minimum velocity of the load motion at which the destruction of ice occurred, was significantly less than in the case of an infinite depth of the bottom and was about u_m=1.55 m/s. When reducing model velocity the area of the fracture was abridged significantly, although the intensity of destruction at low velocity was higher than at a velocity greater than the critical value, which is associated with high curvature of FGW.

Character of the failure of the model ice is shown in figures 19-20. It is should be noted that most intensive ice failure and ice pieces crushing occured at near-critical velocity. Ice cover completely lost its carrying capacity. When reducing model velocity the area of the fracture was abridged significantly, although the intensity of destruction at low velocity was higher than at a velocity greater than the critical value, which is associated with high curvature of FGW.

Figure 19. Character of the failure of the model ice by FGW: h_2=0.3 m (the projection №1); u_m=1.78 m/s

Figure 20. Character of the failure of the model ice by FGW: h_2=0.37 m (the projection №3); u_m=1.62 m/s

4. Conclusion

The results of the studies show that water depth has a significant impact on the parameters of FGW generated by the load motion. Given that the destruction of the ice takes place at velocity much lower than in the deep water. The value of critical velocity also depends on the water depth. Practically in all considered cases an intensive ice pieces crushing occured; and ice cover completely lost its carrying capacity. This study showed that the process of passing the projection of FGW from the ledge leads to a sharp change of its parameters. The velocity of the load motion leading to intensive destruction of modeled ice was far less by comparison with the previously set value for deep water; and the degree of destruction and grinding of ice floes was larger.

References

[1] Kheisin D E 1967 *Dynamics of Ice Cover* (Gidrometeoizdat: Leningrad) p 216
[2] Bukatov A E and Zharkov V V 1995 *Izv. Ross. Akad.Nauk, Mekh. Zhidk. Gaza* vol 2 p 118–125
[3] Kozin V M and Pogorelova A V 2008 *Int. J. Offshore Polar Eng, ISOPE* vol 18 p 271–276
[4] Kozin V M, Onishchuk A G and Mar'in B N 2005 *The Ice-Breaking Capacity of Flexural-Gravity Waves Produced by Motion of Objects* (Dal'nauka: Vladivostok) p 250
[5] Orlov M Yu, Orlova Yu N and Bogomolov G N 2016 *J. Phys. Conf. Series* **774** 012052
[6] Orlov M Y and Orlova Y N 2015 *State University Journal of Mathematics and Mechanics* No 6, p 81-89
[7] Zemlyak V L, Baurin N O and Petrosyan G V 2014 *International Youth Scientific Conference «Current issues of continuum mechanics and celestial mechanics»* (Tomsk) p 40-42
[8] Zemlyak V L, Baurin N O and Kurbackiy D A 2013 vol 12 No 1 p 75-84
[9] Voytkunsky J I 1985 *Handbook of ship theory* (Shipbuilding: St. Petersburg) p 584
[10] Kozin V M, Zemlyak V L 2013 *Physical basis of the ice cover destruction by the resonance method* (Institute of Machining And Metallurgy FEB RAS: Komsomolsk-on-Amur) p 250
[11] Stepanjuk I A 2001 *Test Technology and sea ice modeling* (Gidrometeoizdat: St. Petersburg) p 78
[12] Lewis J W 1982 Recent Development in Physical Ice Modeling *Offshore Technol. Conf. Houston* vol 4 p 493-498

[13] Zemlyak V L, Kozin V M and Baurin N O 2015 *Proc 25th Int Offshore and Polar Eng Conf, ISOPE* p 1884-1889

Acknowledgments
The reported study was funded by RSF (Russian Science Foundation) according to the research project No. 16-19-10097

IOP Conf. Series: Journal of Physics: Conf. Series **919** (2017) 012005 doi:10.1088/1742-6596/919/1/012005

On the theory of multi-pulse vibro-impact mechanisms

L A Igumnov, V S Metrikin, I V Nikiforova and A A Ipatov

Research Institute for Mechanics, National Research Lobachevsky State University of Nizhni Novgorod, 23, bldg. 6, Prospekt Gagarina (Gagarin Avenue), Nizhny Novgorod, 603950, Russia

E-mail: v.s.metrikin@mail.ru

Abstract. This paper presents a mathematical model of a new multi-striker eccentric shock-vibration mechanism with a crank-sliding bar vibration exciter and an arbitrary number of pistons. Analytical solutions for the parameters of the model are obtained to determine the regions of existence of stable periodic motions. Under the assumption of an absolutely inelastic collision of the piston, we derive equations that single out a bifurcational unattainable boundary in the parameter space, which has a countable number of arbitrarily complex stable periodic motions in its neighbourhood. We present results of numerical simulations, which illustrate the existence of periodic and stochastic motions. The methods proposed in this paper for investigating the dynamical characteristics of the new crank-type conrod mechanisms allow practitioners to indicate regions in the parameter space, which allow tuning these mechanisms into the most efficient periodic mode of operation, and to effectively analyze the main changes in their operational regimes when the system parameters are changed.

1. Introduction

Recently, alongside with shock-vibration mechanisms with unbalance vibration exciters, eccentric shock-vibration mechanisms (ESVM) with a crank-sliding bar vibration exciter (CSVE) [1, 2] have come to be widely used in civil engineering. This innovative design was based on the principle of "the inverted vibrator", with its working unit, unbalanced mass, hinged on an eccentric shaft and balanced in rotation by the unbalanced mass. The loading pulse transmitted onto the surface (ground, piles etc.) is produced both due to a spreader with a shoulder of the eccentric shaft and due to the kinetic energy of the falling of the working unit. The effectiveness of compacting and immersing machines not only depends on the quantity of the energy transmitted to the medium being processed, but, to the larger extent, on the character of transmitting this energy – the pulse "form" which should be varied by rearranging certain dynamic factors of a single loading pulse. It is evident that a dense and, at the same time, strong structure of the ground can be achieved only when specific pressure on the surface of the contact of the working unit and the ground in the process of compaction is increased by degrees, its lower limit being determined by the physical properties of the ground in its initial state (before compaction), and the upper one by the ultimate strength of the ground or by technological conditions.

Thus, parameters of such machines and mechanisms must be defined from the conditions close to quasi-plastic interaction. Besides, the pulse rate in each single cycle must be such that it precludes possible development of elastic aftereffects of the processed medium in the intervals between the pulses. Such multi-pulse loading mode can be realized using multi-striker ESVM with CSVE designed to easily regulate working modes by changing the geometry of cinematic connections and to solve the problem of compaction of soils in strained conditions of industrial and civil engineering processes.

The variety of problems that require the study of dynamic models with discontinuous or piecewise-continuous non-linearities is enormous. Thus, nonlinear dynamic problems of vibration systems with impact interaction of the elements are currently being analyzed by scientific teams headed by B.I. Van de Vrande, D.N. Van Campen, P.I. Leine, G.W. Luo, V.K. Astashov , V.I.Babitsky, M.I. Feygin et al [3-18]. These works analyzed the complex dynamics of vibration systems with impact interactions of various nature. Periodical motion modes in non-self-supporting systems with finite (mostly with one) and infinite (quasi-plastic) number of impacts during one period of the periodic motion have been studied. Where possible, the mathematical apparatus of the point mapping method [17-28] was used in the studies. The systems and mechanisms used as vibro-impact rams for various media (soils, concrete, loose materials etc.) as well as for driving in or extracting piles, were designed in the form of unbalance mechanisms. The designs of such mechanisms, especially when being used as vibro-impact rams, often could not cope with their tasks for various reasons. One of such reasons is their being unsuited for working in conditions of limited space, as well as being highly power-consuming.

2. Problem setting

The impact-vibration mechanism considered in the paper (figure 1) consists of the frame 1 encasing the eccentric shaft 2 with flywheels at its ends.

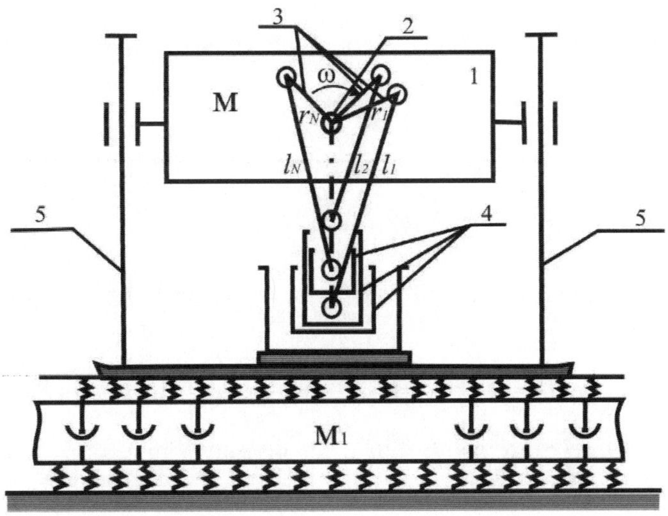

Figure 1. The impact-vibration mechanism.

On the shaft, there are eccentric mechanisms 3, each consisting of two eccentrics nested into each other with a possibility to change the position of the washers, which makes it possible to regulate the value of eccentricity r_i and phase shift φ_i between them (i = 1,2, ..., N). Slides-strikers (SS) 4 are hinged at the free ends of the connecting rods. The eccentric mechanisms in combination with the connecting rods and SS convert the rotary motion of the of the shaft with the constant cyclic rate of rotation of the flywheel into reciprocating motion of the frame relative to stands 5. Each SS strikes its own anvil.

The processed medium (soil, pile, etc.) is represented as an elastically fastened mass M_1 with elasticity coefficient C. Energy losses in the slide-bars of the frame and the medium is accounted for in the form of viscous friction with equivalent damping coefficients b and b_1, respectively.

As the eccentrics can be phase-shifted by angles φ_i relative to each other and have, generally speaking, different lengths, the body oscillates relative the mechanism with the largest sum of projections of the geometrical dimensions onto the vertical axis at the moment. The impact on the anvil of one or several SS's takes place either during the change of SS's interacting with the anvil, or after the breakage of the body together with the eccentric mechanisms.

Neglecting the masses of SS's, connecting rods and cranks, the equation of free (without impact) motion of the system for $y_{pi} > y_c$ can be written in the following form

$$M\frac{d^2y}{dt^2}+b\frac{dy}{dt}=-Mg; M_1\frac{d^2y_c}{dt^2}+b_1\frac{dy_c}{dt}+Cy_c=-M_1g, \tag{1}$$

where y – coordinate of the mass centre of the body counted off from the equilibrium position of mass M_1 for a non-deformed spring C, y_c is deviation value of mass M_1 from the same initial point, y_{pi} are values characterizing deviation of bases of the i–th SS, g is free-fall acceleration.

When one of the SS contacts the medium $y_{pi} = y_c$ there takes place either a momentary impact interaction, such that if $\dot{y}^-_{pi}$, $\dot{y}^-_c$ and $\dot{y}^+_{pi}$, $\dot{y}^+_c$ are velocities of the i-th SS and the medium immediately before and after the impact interaction, respectively, then

$$\begin{aligned}\dot{y}^+_{pi}&=[(\mu_0-R)\dot{y}^-_{pi}+(1+R)\dot{y}^-_c]/(1+\mu_0),\\ \dot{y}^+_c&=[\mu_0(1+R)\dot{y}^-_{pi}+(1-\mu R)\dot{y}^-_c]/(1+\mu_0),\end{aligned} \tag{2}$$

or simultaneous $(R=0)$ motion of the shock-vibration mechanism with the medium described by the following equation:

$$M\frac{d^2y}{dt^2}+b\frac{dy}{dt}+M_1\frac{d^2y_c}{dt^2}+b_1\frac{dy_c}{dt}+Cy_c=-(M+M_1)g \tag{3}$$

In equations (2) the following definitions are used: $0 \le R \le 1$ is impact velocity recovery coefficient, $\mu_0 = M/M_1$.

The position of the eccentricities of length will be measured by angles $\theta_i = \omega t - \varphi_i$, counted off the vertical axis. Then it follows that

$$y_{pi}=y-s_i+r_i\cos(\omega t-\varphi_i)-\sqrt{l_i-r_i^2\sin^2(\omega t-\varphi_i)}, \tag{4}$$

where s_i is distance from the fixture point of the *i*-th connecting rod to the base of the *i*-th SS, l_i is length of the *i*-th connecting rod.

When changing in system (1-4) to dimensionless time $\tau = \omega t$, coordinates $x=(y-s_2-l)/l$, $x_1 = y_c/l$ of the body and the medium, parameters

$$\mu=r_1/l, \gamma_i=r_i/r_1, \varepsilon_i=(s_i-s_2)/l, p=g/\omega^2 l, \lambda^2=C/M_1\omega^2, 2h=b/M\omega, 2h_1=b_1/M_1\omega, l_i=l,$$

as well as introducing function $f_i(\tau)=\varepsilon_i-\mu\gamma_i\cos(\tau-\varphi_i)$, for $r_i \le\le l$ the following equations describing impact-oscillatory motions of the mechanism, accounting for the mobility of the processed medium, are obtained:

$$\begin{cases}x-x_1>f(\tau),\\ \ddot{x}+2h\dot{x}=-p,\\ \ddot{x}_1+2h_1\dot{x}_1+\lambda^2x_1=-p,\end{cases} \tag{5}$$

$$\begin{cases} \dot{x}^{+} = [(\mu_0 - R)\dot{x}^{-} + (1+R)\dot{x}_1^{-} + (1+r-\mu_0)\dfrac{df(\tau)}{d\tau}]/(1+\mu_0), \\ \dot{x}_1^{+} = [\mu_0(1+R)\dot{x}^{-} + (1-R\mu_0)\dot{x}_1^{-} - \mu_0(1+R)\dfrac{df(\tau)}{d\tau}]/(1+\mu_0), \\ (1+\mu_0)\ddot{x} + 2(h_1 + \mu_0 h)\dot{x} + \lambda^2 x = -(1+\mu_0)p + F(\tau), \\ x - x_1 = f(\tau), \end{cases} \quad (6)$$

τ -periodical function $F(\tau)$ on equations (6) describing simultaneous motion of the striking mechanism with the medium after the impact of the *i*-th SS has the following form:

$$F(\tau) = \varepsilon_i + (\lambda^2 - 1)f(\tau) + 2h_1\frac{df(\tau)}{d\tau}, \quad (7)$$

$$f(\tau) = \max\{f_1(\tau), f_2(\tau), ..., f_N(\tau)\}.$$

3. Solution method

Assuming $x_1 = 0$, $M_1 = \infty$, $h = 0$ in equations (5-7), the equations of motion with momentary stops [3] of one or several SS's can be rewritten as

$$\begin{aligned} &\ddot{x} = -p, x > f(\tau), \\ &\dot{x}^{+} = -R\dot{x}^{-} + (1+R)\frac{df(\tau)}{d\tau}, x = f(\tau). \end{aligned} \quad (8)$$

The phase space of system (8) in coordinates $x, \dot{x}, \tau$ is truncated along x, and, besides, $x \geq f(\tau), \dot{x} < +\infty$ (figure 2).

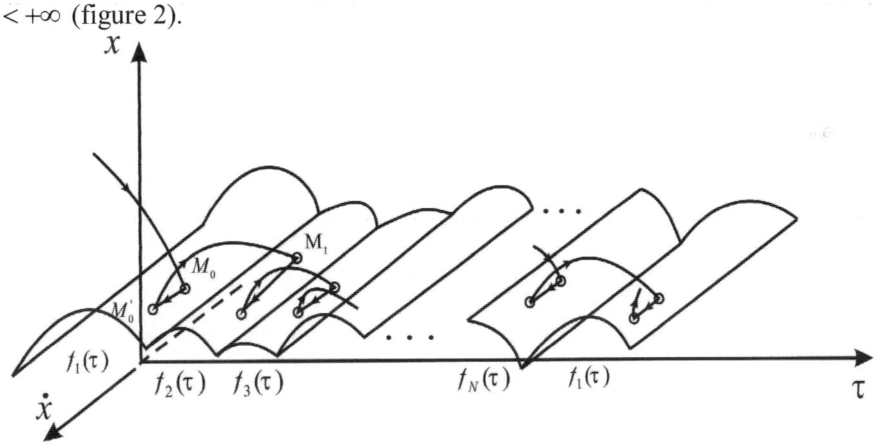

Figure 2. The phase space.

All the phase trajectories are situated either along the surface $x = f(\tau)$ or above it, which represents (generally speaking) N of intersecting cylindrical surfaces $x = f_i(\tau)$ $(i = 1, 2, ..., N)$. The case of $x = f(\tau)$ corresponds to free motion of the falling mechanism, and $x = f(\tau)$ corresponds to impact interaction of one of the SS's with the stopper.

In the case of impact interactions of SS with the stopper for $R \neq 0$, the representing point in the phase space moves as follows: from region $\Phi_1(x > f(\tau))$, the representing point gets on surface $\Pi(x = f(\tau))$ into a point $M_0(\tau_0, x_0 = f(\tau_0), \dot{x}_0)$ (see figure 2), and then, moving along surface Π into point $M_0^1(\tau_0, x_0 = f(\tau_0), \dot{x}_0^{+} = -R\dot{x}_0 + (1+R)df(\tau_0)/d\tau)$, leaves it and gets into subspace Φ_1, where it stays until it gets once again onto surface Π into a point $M_1(\tau_1 > \tau_0, x_1 = f(\tau_1), \dot{x}_1)$, etc. If at a time $\tau = \tau_i$, when the representing point gets onto surface Π into point $M_i(\tau_i, x_i = f(\tau_i), \dot{x}_i)$, the value before the impact velocity is equal to the velocity of change $f(\tau)$ at that moment, or a limiting case of the value of the velocity recovery coefficient is considered for the impact $R = 0$, then the representing point displaces from M_i to point $M_i^1\left(\tau_i, x_i, \dot{x}_i^+ = df(\tau_i)/d\tau\right)$ and then will slide along surface Π until, at time $\tau = \tau_s$, it reaches boundary $\mu\gamma_i \cos(\tau_s - \varphi_i) + p = 0$ of the sliding potion plate. These time intervals $\Delta_{si} = \tau_s - \tau_i$ correspond to prolonged contacts of SS with the stopper.

In what follows, the main attention will be paid to periodical motion modes with successive impacts of each SS during period Γ. It is evident that such a mode is only possible under the condition of paired intersection of two successive surfaces $f_k(\tau), f_{k+1}(\tau)$. For this to be possible, the following relations for the system parameters must hold:

$$(\varepsilon_{k+1} - \varepsilon_k)^2 \le \mu^2(\gamma_{k+1}^2 + \gamma_k^2 - 2\gamma_k\gamma_{k+1}\cos(\varphi_{k+1} - \varphi_k)) \tag{9}$$

If for any pair $f_l(\tau), f_{l+1}(\tau)$ this inequality does not hold (no intersection of the surfaces) then, in the technological process of the motion of the mechanism, the mode with the stroke of the *l*-th or the (*l*+1)-th SS is skipped. To this end, conditions (9) are of paramount importance when adjusting a specific mechanism to the mode with alternating strokes of each of the SS's N.

4. Constructing a point map

It follows from the description of the structure of phase space Φ in system (8) that the dynamics of the mathematical model considered can be investigated by studying the properties of the point transform [18-19,24-28] T of surface Π into itself.

Let us introduce into the analysis the simplest point transforms T_{k+1} transforming surface points $x = f_{k+1}(\tau)$ into surface points $x = f_{k+2}(\tau)$, which, using (8), can be written in the following form $(h = 0)$

$$\begin{aligned} &f_{k+2}(\tau_{k+1}) = \Delta\tau_{k+1}(\dot{x}_k - p\Delta\tau_{k+1}/2) + f_{k+1}(\tau_k), \\ &\dot{x}_{k+1} = R(p\Delta\tau_{k+1} - \dot{x}_k) + (1+R)\frac{df(\tau_{k+1})}{d\tau}, \\ &(\Delta\tau_{k+1} = \tau_{k+1} - \tau_k, k = 0,1,\ldots,N-1). \end{aligned} \tag{10}$$

The parameters in (10) have to satisfy relations (9) as well as the conditions of existence of point transforms T_{k+1}, namely

$$\begin{aligned} &f_{k+1}(\tau_k) \ge f_l(\tau_k), k = 0,1,\ldots,N-1, l = 1,\ldots,N, \\ &x(\tau) > f(\tau), \tau_k < \tau < \tau_{k+1}. \end{aligned} \tag{11}$$

Point transform T of point $M_1 \in f_1(\tau_1)$ into point $M_{n+1} \in f_1(\tau_{n+1})$ under conditions (11) is defined as

$$T = \prod_{i=N}^{1} T_i. \tag{12}$$

5. Existence and stability conditions for fixed points at a point map T corresponding to periodic two-impact motion modes

If S_i is the number of strokes of the *i*-th SS against the stopper during one period $f(\tau)$, and n is multiplicity of period Γ of the periodic motion mode to period $f(\tau)$, then the periodic motion will be characterized by numbers S_i, n $(i = 1,2,...,N)$. It means that the problem of analyzing the periodic motion mode with a single stroke of each of the N SS'*s* can be reduced to analyzing the properties of point mapping T and the conditions of existence of the latter (C - bifurcations), analogous to analyzing systems with discontinuous nonlinearities [18, 19]. Thus, the coordinates of a fixed point $\mathrm{M}^*(\tau^*, \dot{\mathrm{X}}^*)$ corresponding to the periodic motion mode with successive strokes of each of the SS's are determined from system $2(N+1)$ of equations (8) supplemented by periodicity conditions

$$\dot{\mathrm{X}}_{N+1} = \dot{\mathrm{X}}_1 = \dot{\mathrm{X}}^*, \tau_{N+1} = \tau_1 + n\Gamma = \tau^*. \tag{13}$$

Using simple transformations, it can be found from system (8) and (13) that

$$\begin{aligned} \dot{\mathrm{X}}^* &= \frac{b_N - R^N \sum_{k=1}^{N} (-1)^{k+1} b_{N-k}}{1 + (-1)^{N-1} R^N}, \\ \dot{\mathrm{X}}_{k+1} &= R^k [(-1)^{k+1} \dot{\mathrm{X}}^* + \sum_{i=0}^{N} (-1)^i b_{k-i}], \end{aligned} \tag{14}$$

where the components of N-dimensional vector-function $b(b_1, ..., b_N)$ are independent of $\dot{\mathrm{X}}^*, \dot{\mathrm{X}}_{k+1}$ but are functions of τ^*, τ_{k+1} and the parameters of the system

$$b_j = Rp(\tau_{j+1} - \tau_j) + (1+R)\frac{df_{j+1}(\tau_{j+1})}{d\tau}. \tag{15}$$

Times $\tau_k, k = 1,2,3,...N$ are determined by solving the system of nonlinear equations of the form

$$\begin{aligned} & f_{j+1}(\tau^*_{j+1}) + \Delta\tau^*_{j+1}(p\Delta\tau_j / 2 - \dot{\mathrm{X}}^*_j) = f_j(\tau^*_0), \\ & f_1(\tau^*) - f_N(\tau_N) + (\tau^* - \tau^*_N)[p(\tau^* - \tau^*_N)/2 - \dot{\mathrm{X}}^*_N] = 0, \\ & (j = 1,2,...,N-1). \end{aligned} \tag{16}$$

Stability in the small of the main of the periodic motion modes depends on the value of the roots of the characteristic equation

$$\chi(Z) \equiv a_0 Z^2 + a_1 Z + a_2 = 0,$$

where coefficients $a_l, l = 0,1,2$ are defined, as usual [4], after the linearization of the equations of point transform of T in the vicinity of a fixed point. After some simple transformations one obtains

$$\chi(Z) = \begin{pmatrix} A(Z)...B \\ C(Z)...D(Z) \end{pmatrix} = 0,$$

where $A(Z), B, C(Z), D(Z)$ are square matrices, with non-zero entries having the following form:

$$a_{i,j+1} = (-1)^j (p\Delta\tau_{i+1} - \dot{\mathrm{X}}^*_i) + \frac{df_{i+j}(\tau^*_{i+j})}{d\tau},$$

$$a_{N,1} = Z\left[p(\tau^1_N - \tau^*) + \dot{\mathrm{X}}^*_N - \frac{df_1(\tau^*)}{d\tau}\right],$$

$$c_{i,i+j} = -R(-1)^j p + (1+R) j \frac{d^2 f_{i+1}(\tau^*_{i+1})}{d\tau^2},$$

$$c_{N,1} = Z\left[Rp + (1+R)\frac{d^2 f_1(\tau^*)}{d\tau^2}\right],$$
$$b_{i,i} = \Delta\tau^*_{i+1}, d_{i,i+j} = (j-1)R - j, \tag{17}$$
$$d_{N,1} = -Z, j = 0,1; i = 1,2,...,N.$$

Let us consider in a more detail the dynamics of the mechanism for various values of N of SS's.

6. The dynamics of single piston mechanism

$N = 1$. Using (14-17), the coordinates of the fixed point are determined through the parameters of the system as

$$\tau^* = \pi - \arcsin\Omega, \dot{X}^* = \pi np, \tag{18}$$

where $\Omega = \pi np(1-R)/(\mu(1+R)) \leq 1$.

The regions of existence and stability of periodic single-stroke motion modes are limited by surfaces N_+, N_-, N_φ, whose equations in the parametric form have, respectively, the following form

$$\mu_+ = \Omega\mu, \mu_- = p\sqrt{4(1+R^2)^2 + [\pi n(1-R^2)]^2} / (1+R)^2, R_\varphi = 1$$

Figure 3 shows subdivision of the parameter plane $\mu / p, R$ into the planes (between the upper N_+ and the lower N_- boundaries) of existence and stability of a periodic single stroke n- tuple motion mode.

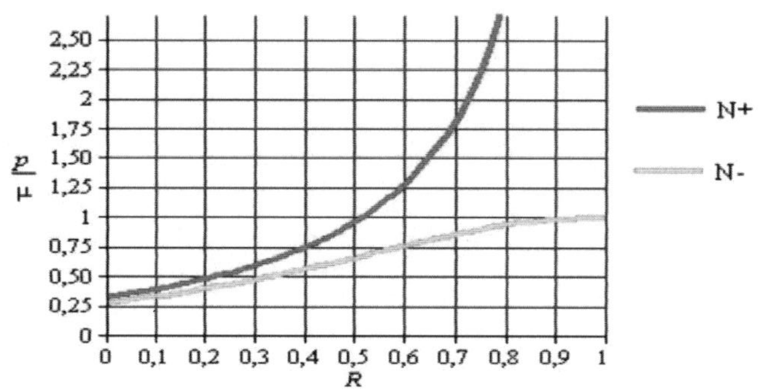

Figure 3. The boundaries of the region of existence and stability.

In Figure 4 a bifurcation diagram for parameter p is shown for and $\mu = 0,12$, wherefrom it follows that a single-stroke motion mode of the mechanism does exist on the interval for the frequency parameter presented in Figure 3. In the region of the parameters that do not belong to this interval, there exist, as was expected, periodic motions with two (Feigenbaum doubling bifurcation [9]), three, etc. strokes of *SS* for the period of motion.

It is to be noted that for $N = 1$, $f(\tau)$ is an analytical function and, thus, by substituting variable $y = x - f(\tau)$ in equations (8), a mathematical model of a conventional impact oscillator is obtained. Its dynamics has been investigated in numerous works (see [4, 5] and the references thereof).

N=2. The phase space of system (8) in coordinates $x, \dot{x}, \tau$ is truncated along x, and, besides, $x \geq f(\tau), \dot{x} < +\infty$ (figure 5).

Conditions (9) necessary for the existence of the motion modes with strokes by each of the two SS's in our case (*N*=2) depict in the parameter plane (figure 6) corresponding regions (shaded parts) for various values of ϕ_i.

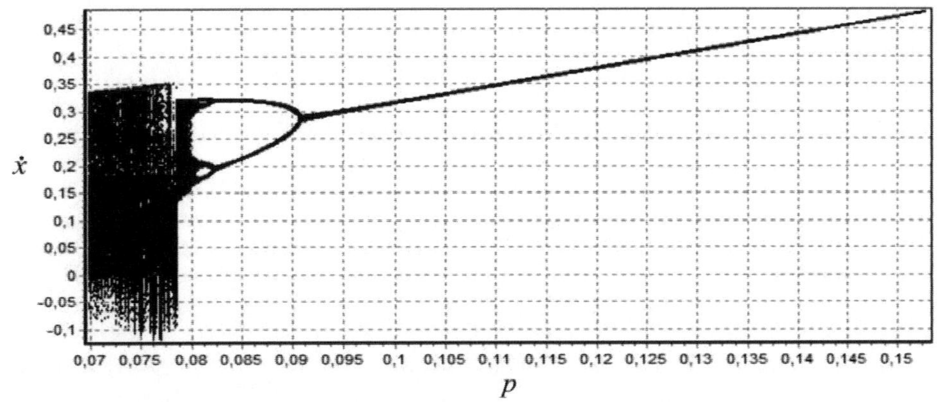

Figure 4. The bifurcation diagram for parameter.

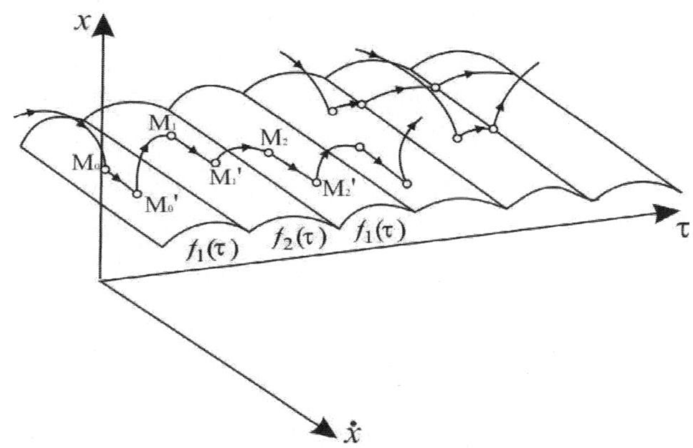

Figure 5. The phase space of system (N=2).

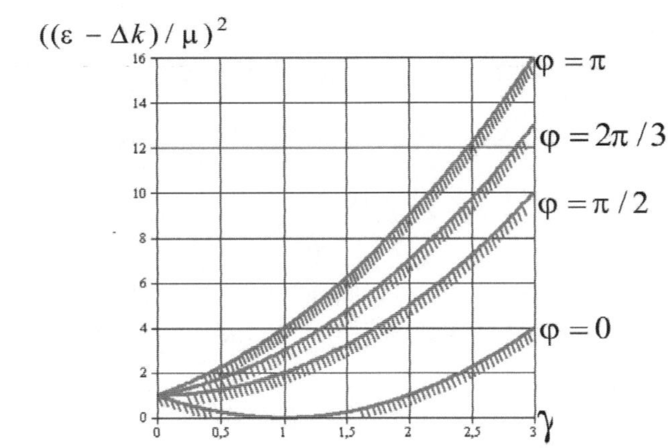

Figure 6. The regions existence of the motion modes with strokes by each of the two SS's.

One can see that with increasing the phase shift between eccentricities r_i the regions of existence of the motion mode with alternating strokes of SS's grow in the parameter space. Point transform T of points

$$M_0(\tau = \tau_0, x_0 = f_1(\tau_0), \dot{x} = \dot{x}_0)) \to M_1(\tau = \tau_1, x_1 = f_2(\tau_1), \dot{x} = \dot{x}_1)) \to M_2(\tau = \tau_2, x = f_1(\tau_2), \dot{x} = \dot{x}_2)$$

can be written as

$$-\mu\gamma\cos(\tau_1-\varphi)=-p(\tau_1-\tau_0)^2/2+\dot{x}_0(\tau_1-\tau_0)+\varepsilon-\mu\cos\tau_0, \tag{19}$$

$$\dot{x}_1=R(p(\tau_1-\tau_0)-\dot{x}_0)+(1+R)\mu\gamma\sin(\tau_1-\phi), \tag{20}$$

$$\varepsilon-\mu\cos\tau_2=-p(\tau_2-\tau_1)^2/2+\dot{x}_1(\tau_2-\tau_1)-\mu\gamma\cos(\tau_1-\varphi), \tag{21}$$

$$\dot{x}_2=R(p(\tau_2-\tau_1)-\dot{x}_1)+(1+R)\mu\sin\tau_2. \tag{22}$$

7. Conditions of existence and stability of fixed points corresponding to periodic two-stroke motions modes.

Equations for determining fixed points of transform T corresponding to two-stroke (with alternating strokes of each SS) periodic motions (the main mode), taking account of the periodicity conditions, can be written in the following way

$$\begin{gathered}\dot{x}_1=R(p\xi-\dot{x})+(1+R)\mu\gamma\sin(\tau_0+\alpha),\\ \dot{x}=R(p(2\pi n-\xi)-\dot{x}_1)+(1+R)\mu\sin\tau_0,\\ -\mu\gamma\cos(\tau_0+\alpha)=-\frac{p}{2}\xi^2+\xi\dot{x}+\varepsilon-\mu\cos\tau_0,\\ \varepsilon-\mu\cos\tau_0=-\frac{p}{2}(2\pi n-\xi)^2+(2\pi n-\xi)\dot{x}_1-\mu\gamma\cos(\tau_0+\alpha),\\ (\alpha=\xi-\varphi,\ \xi=\tau_1-\tau_0).\end{gathered} \tag{23}$$

After a number of transformations (23) can be rewritten as

$$\mu=\frac{\sqrt{(Ad-bB)^2+(Ac-Ba)^2}}{|ad-bc|},$$

where

$$\begin{gathered}a=(1-R)(1-\gamma\cos\alpha)+R\gamma\xi\sin\alpha,b=\xi(1-R\gamma\cos\alpha)-(1-R)\gamma\sin\alpha,\\ c=(1-R)(1-\gamma\cos\alpha)+(2\pi n-\xi)\gamma\sin\alpha,d=(2\pi n-\xi)(R-\gamma\cos\alpha)-(1-R)\gamma\sin\alpha\\ A=(1-R)(\varepsilon-\Delta k-\eta)+\frac{p\xi}{(1+R)}\left(2\pi nR-\frac{\xi}{2}(1+R)^2\right)\\ B=(1-R)(\varepsilon-\Delta k-\eta)+\frac{p(2\pi n-\xi)}{2(1+R)}\left((2\pi n-\xi)(1+R^2)-2R\xi\right).\end{gathered}$$

Stability of a fixed point is determined, as is known [24-28], by the value of the roots of the characteristic equation $\chi(z)=0$, which, in our case, has the following form:

$$\chi(z)=b_{11}b_{22}z^2+(a_{11}b_{22}+(a_{11}b_{22}+a_{22}b_{11}-a_{21}b_{12})z+a_{11}a_{22}-a_{12}a_{21}). \tag{24}$$

The following notations are used in (24):

$$a_{11}=R(1+R)\mu\gamma\cos(\tau_0+\alpha)+Rp+\frac{R^2}{\xi}(\dot{x}-\mu\gamma\sin(\tau_0+\alpha)),$$

$$a_{12}=\frac{R^2}{\xi}(\mu\sin\tau_0-\dot{x}),$$

$$a_{21} = (2\pi n - \xi)\frac{R}{\xi}\left(\dot{x} - \mu\gamma\sin(\tau_0 + \alpha)\right) + (2\pi n - \xi)((1+R)\mu\gamma\cos(\tau_0 + \alpha) + p) - \dot{x}_1 + \mu\gamma\sin(\tau_0 + \alpha),$$

$$a_{22} = (2\pi n - \xi)\frac{R}{\xi}(\mu\sin\tau_0 - \dot{x}), b_{11} = \frac{1}{\xi}(p\xi - \dot{x} + \mu\gamma\sin(\tau_0 + \alpha)),$$

$$b_{12} = \frac{1}{\xi}(\dot{x} - p\xi - \mu\sin\tau_0) - (Rp + (1+R)\mu\cos\tau_0), b_{22} = \dot{x}_1 - (2\pi n - \xi)p - \mu\sin\tau_0.$$

8. The results of the numerical experiments.

Due to the fact that the investigation of complex periodic and stochastic motion modes in the analytical form is a rather complicated and, currently, unworkable problem, the dynamics of such motions was investigated in the present work using numerical experiments on a personal computer with the help of a software complex developed in Borland C++ Builder 6 medium. Figure 7 presents, in plane (p,R), the region of existence and stability (shaded region) of a 2π - periodic motion with alternating strokes of piston-strikers for the following values of the parameters $(\mu = 0.09, \varepsilon = 0.018, \varphi = 3.14, \gamma = 3) \subset P(0)$. Boundary N_c [18] (of the existence of the above described point transform T) cuts from the stability region, bounded by surfaces N_+, N_-, N_ϕ, a small part.

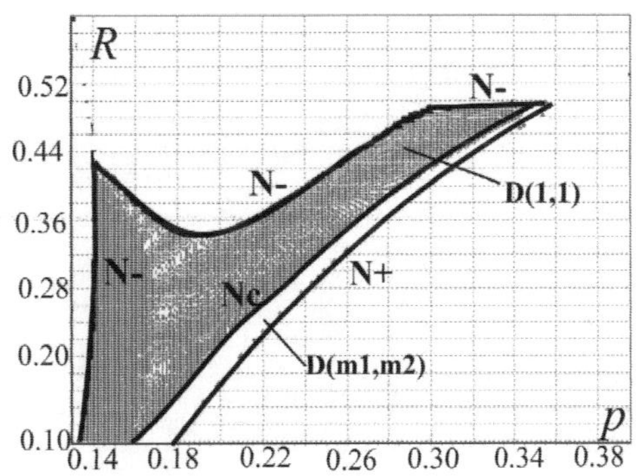

Figure 7. The region of existence and stability (shaded region) of a - periodic motion.

Figure 8 shows a bifurcation diagram for parameter p, constructed for the same parameter values and for R = 0.55. Left part of figure 8 corresponds to the diagram with the impacts on surface $x = f_1(\tau)$; right part of figure 8 – to the diagram with the impacts on surface $x = f_2(\tau)$. It is evident from the figure that for $0.22 \le p \le 0.23$. Doubling the number of strokes; the main motion mode exists in the limits of $0.23 \le p \le 0.25$. For $0.19 \le p \le 0.21$, a chaotic motion mode is observed.

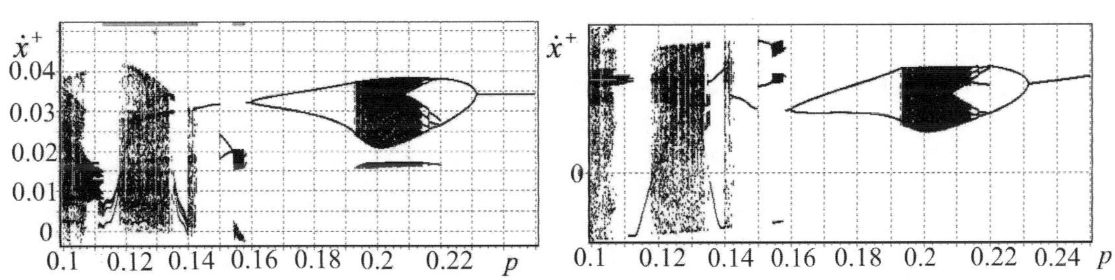

Figure 8. The bifurcation diagrams.

9. Conclusions

IOP Conf. Series: Journal of Physics: Conf. Series **919** (2017) 012005 doi:10.1088/1742-6596/919/1/012005

The work describes:

- a model of a new multi-striker eccentric shock-vibration mechanism with a crank-sliding bar vibration exciter and an arbitrary number of pistons;
- the phase space of the mathematical model;
- equations of a general (for a random number of SS) point mapping of a two-dimensional non-analytical intersecting surface into itself;
- equations (in the parametric form) of the coordinates of the fixed points corresponding to periodic motions with alternating strokes of each of the SS's;
- bifurcation boundaries of existence of the above mentioned equations of point mappings.

The main results achieved:

- a methodology and algorithms of constructing point mapping equations of non-analytical secant surfaces truncated along phase coordinates have been developed;
- the parameter space has been subdivided into regions of stability of periodic motion modes;
- the study of dynamics of a single-piston mechanism undergoing a perfectly inelastic collision has been formulated as a Poincare map of a circle into itself;
- the existence of a denumerable number of stable periodic modes of motion in the neighborhood of unreachable boundaries in the extended phase space has been proven;
- bifurcation values of the parameters have been demonstrated, for which chaotic motion modes exist. The scenario of the appearance of chaotic motion modes is very similar to the scenario of doubling of Feigenbaum period [23-28].

Acknowledgements
The work was financially supported by Russian Scientific Foundation (grant 16-19-10237).

References

[1] Bobylev L M 1980 *Compaction of soils of backfilling in strained working conditions* (Moscow:Stroyizdat)
[2] A.c. 1020479 USSR, A Vibro-ram / Shilkov V.A., Savalyuk A.D., Metrikin V.S., Polyakov A.A., Shabardin A.K., Alyokhin A.I., Omenenko I.Ya. (USSR). - №3376593/29-33; submitted 05.01.82; published 30.05.83, Bull. № 20.
[3] Kobrinskiy A A and Kobrinskiy A E 1973 *Vibro-impact systems* (Moscow: Science)
[4] Babitskiy V I 1978 *The Theory of Vibro-Impact Systems* (Moscow: Science)
[5] Babitskiy V I and Krupenin V L 1985 *Oscillations in Strongly Nonlinear Systems* (Moscow: Science)
[6] Biderman V L 1972 *The applied theory of mechanical oscillations* (Miscow: High School)
[7] Bogodukhov S I, Grebenyuk V F and Proskurin A D 2005 *Processing of Strengthened Surfaces in Mechanical Engineering and Maintenance* (Moscow: Mechanical Engineering)
[8] Vagapov I K, Ganiyev M M and Shinkaryov A S 2008 *Proceedings of Higher Educational Institutions. Machine Building* **5** 3-25
[9] Zakrzhevskiy M V 1980 *Oscillations of a substantially nonlinear substantially-nonlinear mechanical systems* (Riga: Zinatne)
[10] Babitsky V I 1998 *Theory of Vibro-Impact Systems* (New York: Springer)
[11] Cveticanin L 2002 *J. Sound Vib.* **252** 361-369
[12] Pavloaskaia E and Wiercigroch M 2003 *J. Sound Vib.* **267** 893-911
[13] Vagapov I K, Ganiev M M and Shinkarev A S 2007 *J. Sound Vib.* **302** 425-441
[14] Luo G W, Lv X H and Ma L 2008 *Eur. J. Mech. A-Solid.* **27** 1088-1107
[15] Leine R I and Heimsch T F 2012 *Physica* D **241** 2029-41
[16] Goebel R, Sanfelice R G and Teel A R 2008 *Syst. Control Lett.* **57** 980-986
[17] Zheleztsov N A 1949 *J. Appl. Math. Mech.* **13** 3-40
[18] Feygin M I 1994 *Forced oscillations of systems with discontinuous nonlinearities* (Moscow: Science)

[19] Bespalova L V 1957 *News of Acad. Sci. USSR. OTN. Mechanics and Mechanical Engineering* **5** 1-14
[20] Bykhovsky I I, Dorokhova A D, Zaretskiy L B *and* Lukomskiy S I 1964 *News of Acad. Sci. USSR. OTN. Mechanics and Mechanical Engineering* **2** 161-165
[21] Butenina N N and Metrikin V S 1981 *Differential and integral equations* 80-84
[22] Butenina N N and Metrikin V S 1986 *Proceedings of Higher Educational Institutions. Machine Building* **6** 24-28
[23] Shuster G 1988 *Deterministic chaos* (Moscow: World)
[24] AnishhenkoV S and Vadivasova T E 2011 *Lecture Notes on Nonlinear Dynamics* (Moscow-Izhevsk: NIC "Regular and Chaotic Dynamics")
[25] Nekorkin V I 2012 *Lectures on the foundations on the theory of oscillations* (Nizhny Novgorod: UNN)
[26] Neimark Yu I 2012 *Dynamical Systems and Controllable Processes. Second Edition* (Moscow: Librokom)
[27] Kuznecov A P, Savin A V and Tjurjukina L V 2010 *Introduction to physics of nonlinear maps* (Saratov: Nauchnaja kniga)
[28] Danilov Ju A 2006 *Lecture Notes on Nonlinear Dynamics: Basic Introduction. Second Edition* (Moscow: KomKniga)

IOP Conf. Series: Journal of Physics: Conf. Series **919** (2017) 012006 doi:10.1088/1742-6596/919/1/012006

Research of the behavior of ice on water under explosive loads

M Yu Orlov[2], Yu Orlova[2], G Bogomolov[1] and V Glazyrin[1]

[1]National Research Tomsk State University, 634050 Tomsk, Russia
[2]National Research Tomsk Polytechnic University, 634050, Tomsk, Russia

E-mail: orloff_m@mail.ru

Abstract. Complex theoretical and experimental studies of the behavior of ice under dynamic loads are presented. New experimental data about blast of snow-covered ice have been obtained. A complete coincidence of the experimental data of 2013 and 2017 was found. Test calculations, including impact test have been carried out. Shock adiabatics of ice and water are presented. The behavior of the ice block on the water under explosion was investigated. The velocity of the free surface, the volume of ice destruction, the diameter of the explosive lane in ice cover were calculated. Recommendations for more effective ice destruction under these conditions were given.

1. Introduction

The processes occurring in sea and river ice under dynamic loading are the subject of scientific research. This is due to various reasons, including the need to extract natural resources and design transport hubs in the regions of permafrost and the Far North, the development of new icecomposites, the fight against ice jams on Siberian rivers, as well as some military applications. However, the main difficulty lies in the fact that the ice is poorly studied natural material under impact and explosion loading. Ice has unique plastic and dielectric properties, multiple phase transitions in the process of deformation, and in terms of destruction, generally may not have analogues [1].

There is known fact that various scientific teams both in Russia and abroad are engaging the study of the behavior of ice under dynamic load. Systematic scientific research is carried out at Tomsk State University, Far Eastern Federal University, Novosibirsk State Technical University, Lobachevsky University, Krylov Scientific Center, etc. The recognized scientific world center is the Ice Research Laboratory in Germany. However, current research projects about dynamic loading of ice are undetected by the authors [2].

In the current study, complex theoretical and experimental studies of the ice behavior under shock and explosive loads were carried out. In experimental part of the research consists of full-scale underwater explosion tests only. Blast of emulsion explosive Emulast AS-FP (EE) is produced in water under ice. The "Ice-Explosive-Water" system was considered in theoretical part of the research. The behavior ice block on water under explosive was modeled in 2D axisymmetric formulation using an original computer code.

2 Experimental studies of the behavior of ice under dynamic loads

At present time, there are few experimental studies on the deformation and destruction of ice under shock and explosive loads. The overwhelming majority of studies are devoted to the behavior of ice under static and quasistatic conditions. Brief review of this topic is available in [3]. Experimental studies of the explosion of ice were not detected by the authors. Apparently the research has already become a bibliographic rarity.

The response of ice to shock loads was investigated [4-6]. The impact of ice cylinder on rigid wall is considered in [5]. A similar experiment was carried out in [6]. However, the dimensions of the ice cylinder and the initial speed were slightly different. In both cases, the ice was destroyed by impact loads. In research [4] the impact of an ice cylinder on a thin aluminum barrier was considered.

2.1. Mobile laboratory "Explosive Destruction of Natural Materials"

In 2013 the mobile laboratory (mobilab) was organized in National Research Tomsk State University. Currently, mobilab develops as alternative to the resumed American research program "ScICExe". Ice research is conducted with the support of the Ministry of Emergency Situations of Russia and the company KuzbasSpetsVzryv. During this time, the traditional objects of research were natural limestone and river ice of different types. The shape of the explosive crater in the limestone after the blast was found [7]. The diameter and shape of the lane in ice cover were identified after the blast up explosives [8]. Later, the destruction of needle ice and ice cover of sandwich structures "Snow–Shuga – Ice" were studied under the same conditions [9]. The main features of mobilab are carrying out full-scale tests with a maximum explosion about 100 kg of trinitrotoluene (TNT). The main results the author's full-scale tests are the sizes of explosive craters and lanes, the revealed state of the edge of the craters and the morphology of the fragments, etc. Until now, this is treated as quality test.

2.2. Full-scale underwater explosion tests

This section focuses to full-scale underwater explosion tests (UNDEX). Emulsion explosives (EE) were used in all experiments. EE detonation velocity is 4600 m/s. EE charge was placed under ice in freshwater. The charge had a cylindrical shape and a mass of 4 kg (TNT equivalent 3.25 kg). At the moment of the explosion, EE was located horizontally to the ice cover. Water and air temperature was 4 °C. The depth of the water under the ice cover was approximately 5 meters (hydroimpact was excluded). The initiation point of explosion was at the top of charge EE.

2.3. Explosion of ice by EE

This section discusses the results of full-scale tests, with research object was freshwater ice of medium thickness only. The thickness of the ice was in the range from 70 cm to 90 cm. The average thickness of the snow was 15-20 cm. In rare cases, the thickness of the snow layer was 80 cm. In the current research, the results of the blasting of such ice are not given. There were no polynya, crack and fracture in the experimental site (terminology is used in accordance with the Sea Ice nomenclature of 1974). A hole was made in the ice cover to place an explosive in the water. Hole diameter was 11-16 cm.

The subject of the study was the state of the ice cover after the explosion, including the diameter of lane and ice edge inside lane. The diameter and height of the fragment scattering and the size of the fragments were also investigated. Five types of ice were studied by the authors. They are listed below in chronological order: snow covered ice (2013), bar ice (2014), ice cover sandwich structure “Snow - Shuga - Ice” (2015), needle ice (2016) and snow-covered ice (2017).

IOP Conf. Series: Journal of Physics: Conf. Series **919** (2017) 012006 doi:10.1088/1742-6596/919/1/012006

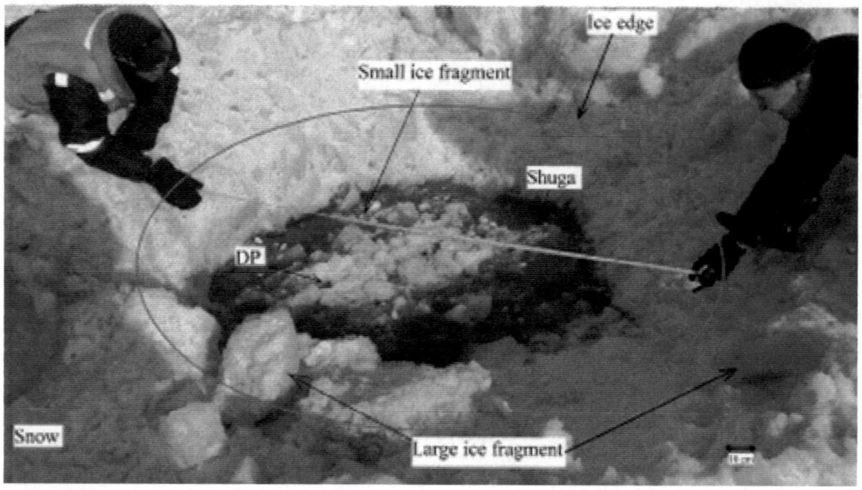

Figure 1. Lane in snow-covered ice cover. Full-scale test 2017. Bar represent 10 cm. Photo by S. Kul'kov

Figure 1 shows the lane after the explosion 4 kg EE. This is the result of full-tests this year. The diameter of the lane was approximately 200 cm. A detailed inspection showed that the lane had a round shape. The photo shows fragments of ice, including large ice fragment (about 50 cm). Inside the lane, small fragments of ice (about 10 cm), shuga and remnants of detonation products (DP) were found. Many remnants of the DP were dispersed by the wind. Under the action of DP, ice fragments and snow were raised up to a height of 10 meters. The ice edge was developed, i.e. smooth and not stepped. The first results of full-scale tests are demonstrated in the author's video [10]. A detailed analysis of the experimental studies revealed that the results of full-scale test of 2013 and the results of the full-scale tests of 2017 were the same.

3. Mathematical model

The ice is described by a phenomenological macroscopic model of the continuum mechanics. The ice model is based on the fundamental conservation laws. Ice model is porous, compressible, strength, with shock-wave phenomena. The Prandtl-Reis equations associated with von Mises equations are used. The shear modulus G and the yield strength σ are not constants. Components of the deviator of the tensor of elastic-plastic stresses are taken from [11].

Polycrystalline ice without phase transitions with averaged mechanical properties is considered. The effect of gravitational forces on ice and other materials is not taken into account. DP action is described by the Landau-Staniukovich polytropic. The appearance of new free surfaces in the process of deformation and destruction is possible. The equation of the state of ice is taken in the simplest form [12].

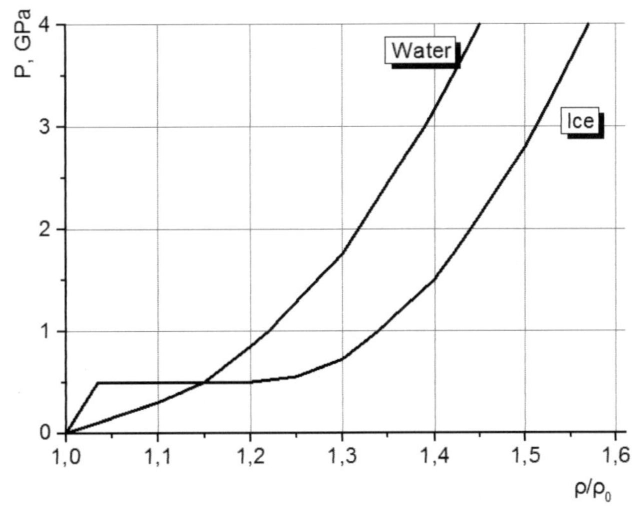

Figure 2. Shock adiabatic of ice and water

The shock adiabats of ice and water is shown in figure 2. The intersection point of the two curves was noted at a pressure of 0.5 GPa. After a pressure of 1 GPa, the curves become parallel.

4. Numerical method

The calculations were carried out in a two-dimensional axisymmetric formulation using a modified lagrangian method. The numerical method contains algorithms for calculating contact surfaces, an algorithm element erosion, an improved algorithm calculating contact surfaces, an algorithm splitting nodes, and so on. Together, algorithms are called a new way of isolating the surfaces of discontinuity in the continuity of materials (According V. Fomin's terminology). The proposed way does not impose serious limitations on the solution of the dynamical multi-contact problems of the mechanics solid body [13].

4.1. Test calculations

To test the mathematical model and the numerical method of calculations, some test calculations were solved. First, the impact test was solved. The impact test consisted in modeling the penetration of a spherical steel impactor into ice block. The impactor velocity varied from 100 m/s to 150 m/s. The post-penetration analysis confirmed that the impact crater in the ice had conical shape. A laboratory experiment was conducted at the Krylov Scientific Center. The discrepancy between the experiment and the calculation was about 8%.

Next, an underwater explosion of the ice cover of 2013 is modeled. The snow cover on the ice was absent in the calculations. The finite element model is constructed by the computer code "Impact 2D". The number of finite elements was 10000. At "Ice – Water" interface, condition of sliding is considered. After this time, the pictures of destruction in the ice remained practically unchanged. According to the calculated data, the explosion of ice was accompanied by the separation of the spall plate. The speed of the spall plate, the time of the initiation of cracks, the volume of destruction and the diameter of the explosive lane were obtained numerically. The discrepancy with the experiment did not exceed 10% [14].

Test calculations have made it possible to determine the strength characteristics of ice, including dynamic yield strength, etc. Physical and mechanical properties of ice and other materials are listed in Table 1. The first column shows the properties of EE and TNT, respectively.

Table 1 Physical and mechanical properties materials

	Emulast AS-FP/TNT	Water	Freshwater Ice
Initial density, g/sm^3	1,34/1,66	1	0,92
Volume velocity of sound, m/s	2540	1500	3020
Shear modulus, GPa	26,9	0,1	3,2
Yield strength, GPa	0,025	0,0001	0, 0022
Spall strength, GPa	0,3	0,001	0,003
Specific work of shear plastic deformations, kJ/kg	100	500	5

5. Numerical research

In this section, numerical simulation of explosive ice destruction is presented. "Ice – Explosive – Water" system was considered. The mechanical properties of ice are given in table 1. The block ice thickness and depth of water were equal to 250 cm. The explosive was laid at a depth of 22 to 221 cm in ice. The charge weight was 4.8 kg. Only nonenveloped charge of TNT was used. TNT height is 23 cm and TNT thickness is 21 cm. The explosive charge was located in the center of the ice block. The number of calculated elements is approximately 20000. The variants of the calculation differed only in the depth of the explosive laying. There were 8 variants of calculations. The process of explosive loading was calculated up to 5 ms.

In the contact zone "Ice - Explosive" the first foci of destruction are formed irrespective of the depth of the explosive laying. Under the action of detonation products, zone of destruction extends from the epicenter of the explosion to the lateral surfaces of the ice block. The formed main cracks are oriented at different angles to the axis of symmetry. Fragmentary destruction of ice was accompanied by splashes of pieces of ice outside through the lane. In the variants 3, 4 and 5, the main crack divided the ice block into two parts.

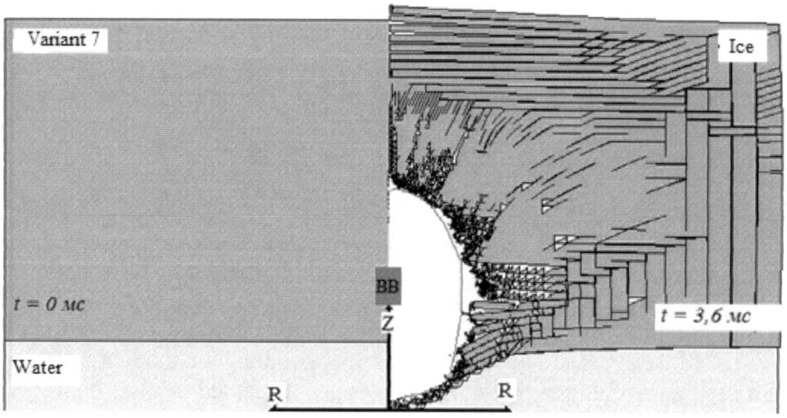

Figure 3. Configurations “Ice - Explosion – Water”, variant 7.

Figure 3 shows the configuration of the "Ice – Explosive - Water" for variant 7. The right side of the figure corresponds to time point of 3.6 ms, but left side of the figure corresponds to time point of 0 ms. The figure illustrates foci of destruction, main cracks and lane after explosive. At this time, the lane is still being formed. Destruction of the side and the back surface has already occurred.

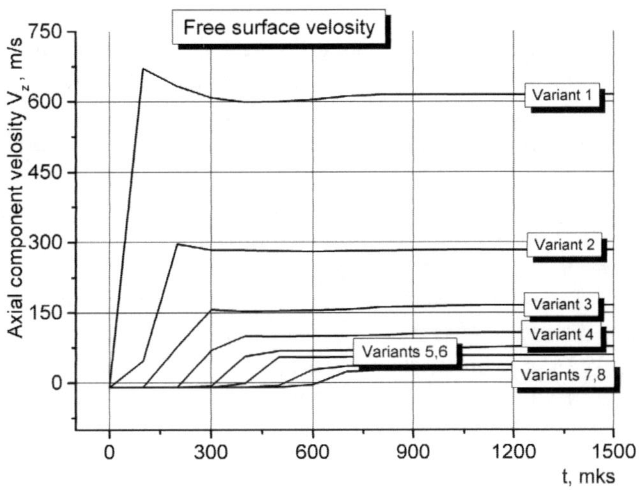

Figure 4. Profiles free surface velocity

Figure 4 shows the free surface velocity profiles for all variants. Velocity calculation performed in the control point on the symmetry axis. Obviously, the maximum velocity was in the first variant of the calculation, and the minimum velocity was in the last variant of the calculation. In the variants of calculations 5 and 6, and also variants 7 and 8, the velocities of the free surface were practically the same. The output of the shock wave on the free surface of the ice block was fixed in the range from 150 to 650 μs. In the range from 650 μs to 1500 ms, the free surface velocity was constant for all variants.

Table 2 presents the calculated parameters of the ice destruction process, including the damage to ice, the diameter of the lane in the ice and the velocity of the free surface.

Table 2. Numerical data of destruction of ice under explosive loads

Variant number	The depth of the explosives [cm]	Ice damage [%]	lane diameter in ice [cm]	Free surface speed of ice [m/s]
Variant 1	22	30,7	110	614
Variant 2	45	33,2	130	282
Variant 3	75	36,8	135	163
Variant 4	102	38,3	137	105
Variant 5	130	36,9	140	77
Variant 6	153	37,2	135	53
Variant 7	193	30,9	128	41
Variant 8	221	26,9	150	32

As a measure of ice damage, a quantitative criterion of materials damage was chosen. The criterion does not contain physical quantities and is bound only to the grid [11]. Maximum volume of ice damage did not exceed 40%. In variants 1-6, an increase in ice damage was noted. In variants 7 and 8, on the contrary, the damage to the ice decreased. The maximum diameter of the lane is fixed in the last variant. There the explosive was in contact with the water. The speed of the free surface decreased with increasing depth of the explosive laid.

6. Conclusion

1. The results of full-scale tests of 2017 coincided with the results of full-scale tests of 2013. Dimensions of lane in ice cover, its shape, ice edge, and morphology of ice destruction were subjects of comparison. From which it follows that the strength properties of these ice are also the same. Mechanical properties of such ice, including dynamic yield strength and other strength characteristics are defined in [1].
2. The process of explosive loading of an ice block on water has been investigated numerically. The velocity of the free ice surface, the volume of ice destruction and the diameter of the explosive lane in the ice were calculated. Depending on the location of the explosives depth free surface velocity varied from 614 m/s to 32 m/s. According to the calculated data, the diameter of the lane varied from 110 cm to 150 cm. The volume of destruction in the ice varied from 26% to 40%. The lower estimate is approximate. In all cases, it was a lot.
   For more effective destruction of ice, it is recommended to lay explosives in the middle of the ice block.

References

[1] Bogorodsky V V and Gavrilo V P 1980 *Ice. Physical properties. Modern methods of glaciology* (Leningrad: Gidrometeoizdat Press) p 383 (in Russian)
[2] http://engineering.dartmouth.edu/icelab/
[3] Orlova Y N 2014 Comprehensive theoretical and experimental study of the behavior of ice under impact and, explosive loads (National Research Tomsk State University) Ph.D. thesis
[4] Combescure A, Cyuzel-Marmot Y 2011 *Int. J. of Solids and Structures,* **48** 2779-2790
[5] Carney K S (Eds) 2006 *Int. J. of Solids and Structures*, **43** 7820–7839
[6] Pernas-Sanchez J, Arterro-Guerrero J, Varas D and Lopez-Puento J 2013 *Proceeding 7th International Workshop on Dynamic behavoir of materials and its application in industrial processes (Madrid)* Spain 47-49
[7] Orlov M Yu, Orlova Yu N and Bogomolov G N 2016 *J. Phys. Conf. Series* **774** 012052
[8] Orlov M Y, Orlova Yu N and Tolkachev V F 2015 *J. Phys.: Conf. Ser.* **653** 012038
[9] Orlov M Y 2017 *Proceeding of Tomsk State University. A series of Physics and Mathematics.* **300** 8-16
[10] Orlov M Yu and Orlov Yu N 2016 *Problems of the Arctic and Antarctic* **1** 28-38
[11] Glazyrin V P 2008 Deformation and destruction of inhomogeneous materials and structure at shock and explosion (National Research Tomsk State University) Ph.D. thesis
[12] Tolkachev V F and Trushkov V G 1993 *Russian Journal of Physical Chemistry B: Focus on Physics.* **35** 170-174
[13] Gerasimov A V, Glazyrin V P i dr. 2007 Theoretical and experimental studies of high-speed interaction of bodies *Tomsk Univ Tom. University Press* 572 (in Russian)
[14] Orlov M Yu and Bogomolov G N 2016 Study of the behavior of ice under shock and explosive loading *Polar Mechanics,* ISBN 9757444-3795-4

IOP Conf. Series: Journal of Physics: Conf. Series **919** (2017) 012007 doi:10.1088/1742-6596/919/1/012007

The efficiency of ceramic-faced metal targets at high-velocity impact

V F Tolkachev, A A Konyaev and N V Pakhnutova

National Research Tomsk State University, 36, Lenin Ave., Tomsk, 634050, Russia

E-mail: tolk@niipmm.tsu.ru

Abstract. The paper represents experimental results and engineering evaluation concerning the efficiency of composite materials to be used as an additional protection during the high-velocity interaction of a tungsten rod with a target in the velocity range of 1...5 km/s. The main parameter that characterizes the high-velocity interaction of a projectile with a layered target is the penetration depth. Experimental data, numerical simulation and engineering evaluation by modified models are used to determine the penetration depth. Boron carbide, aluminum oxide, and aluminum nickelide are applied as a front surface of targets. Based on experimental data and numerical simulation, the main characteristics of ceramics are determined, which allows composite materials to be effectively used as additional elements of protection.

1. Introduction

Different types of ceramics and composite materials (cermets, nanostructured materials) with high dynamic strength characteristics are used as additional elements of protective constructions [1]. To select the most effective ceramics for layered constructions from available and newly developed materials, there is a need in the comparison of shock resistance by varying the initial conditions for the high-velocity impact of compact and rod projectiles. Experimental data for the penetration depth of high-density rods into semi-infinite targets are given for aluminum oxide Al_2O_3, boron carbide B_4C and aluminum nitride AlN in [2, 3]. Reliable methods for the evaluation of characteristics during high-velocity impact are experimental investigation, physical and numerical simulation for the penetration of projectiles into targets and constructions containing the layers of metals, ceramics, and composites [4-11].

The determination of the penetration deep and residual velocity of a projectile behind a target is a very effective test to detect the shock resistance of ceramic materials [2, 5, 7, 8]. In experiments, targets consist of semi-infinite ceramic blocks or configurations of layers, where ceramics is placed on the front surface of semi-infinite targets or between metal plates.

2. Formulation of the problem and investigation results

Based on the experimental and numerical results obtained [5-11], ceramic and composite materials such as aluminum oxide Al_2O_3 (corundum), boron carbide B_4C, and a composite AlN were investigated as additional elements of armor protection, considering strength and weight-dimension characteristics.

The main physical and mechanical characteristics of materials are given in Table 1, where: ρ_0 is the density; E is the Young's modulus; H_M is the microhardness; H_D is the dynamic hardness; σ_T is the yield strength.

Table 1. Physical and mechanical characteristics of materials

Material	ρ_0,10^3, kg/m³	E, GPa	H_M, GPa	H_D, GPa	σ_T, GPa
B_4C	2,4	475	38	42-49	64.5
Al_2O_3	3,8	407	28	20-25	59.3
AlN	3,25		18	16-22	

An imitation model [5, 8] based on the hydro-dynamical model of armour-penetration was proposed to predict the penetration depth of rod projectiles into solid targets. The main advantage of this model is to take into account the dynamic compressibility of a projectile material in the area of contact with a target. The calculations conducted for the penetration depth of tungsten rod projectiles into steel targets at the impact velocities of 1000...5000 m/s are in good agreement with experimental data [2, 3, 5, 6, 8]. Figure 1 shows the relative penetration depth of TNI-90 projectiles (tungsten – 90%, nickel – 3%, iron – 7%) with a length of 6 cm and an elongation of 10 calibers into steel and ceramic semi-infinite targets. The calculation results are in good agreement with experimental data.

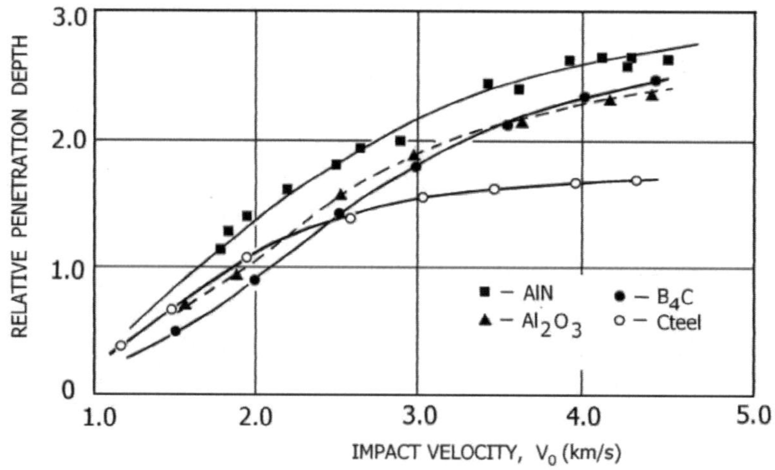

Figure 1. Relative penetration depth of tungsten rods into semi-infinite targets

During shock-wave loading, ceramic plates are broken into fragments and, in most cases, can not be reconstructed after experiments, which does not allow the penetration depth of projectiles to be determined. This fact is illustrated by the X-ray image of the projectile and fragments of the fractured ceramic plate Al_2O_3 at the impact velocity of 760 m/s (figure 2).

The experiments have shown that ceramics has a significant resistance to the penetration of a projectile at impact velocities below 1000 m/s due to high strength characteristics. At the first stage of penetration this fact is proved by the intensive plastic deformation of the head part of the projectile and fragmentation of ceramics in the contact zone. Then the deformed projectile penetrates into the fractured ceramic sample.

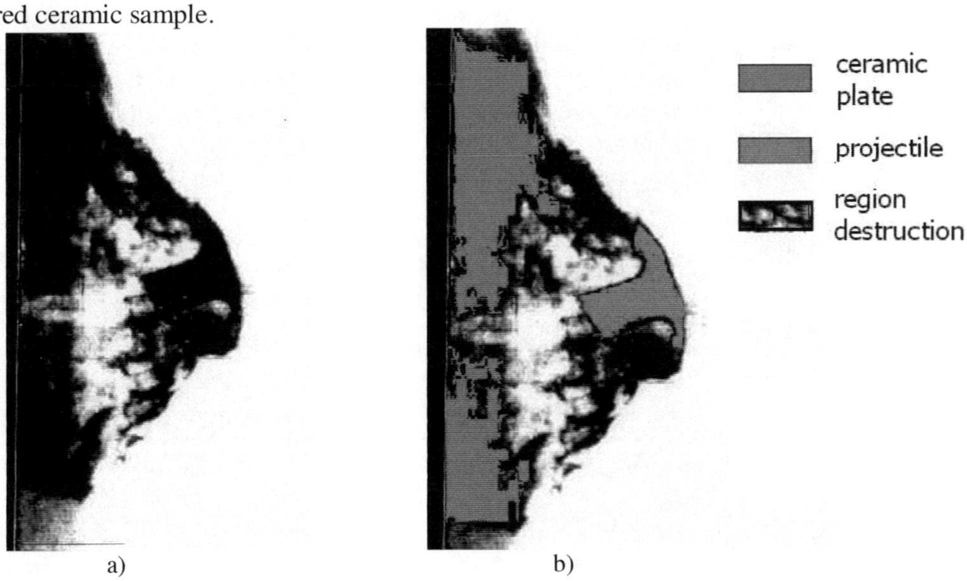

Figure 2. X-ray image of Al_2O_3 ceramics fracture during the impact of a 7.62 mm bullet (a) and the computation for the level distribution of the fractured material (b)

The efficiency of ceramic samples was determined using an indirect method, in particular, by the penetration depth of a projectile into a massive metal target placed behind ceramics according to the scheme, as showed in figure 3.

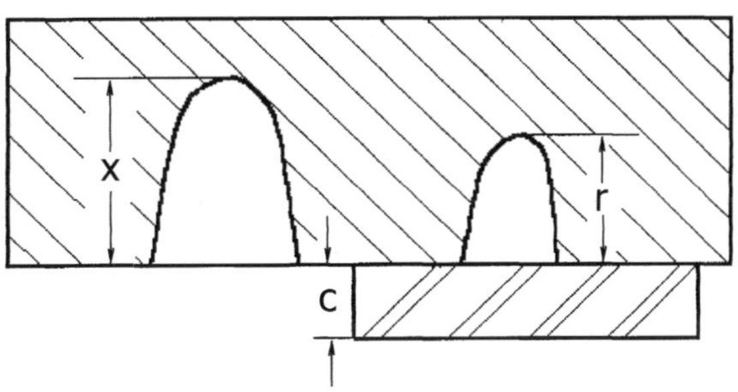

Figure 3. Scheme for recording the penetration depth of a projectile into layered targets

The efficiency of the shock resistance of ceramics (ε) is determined from equation [12]:

$$\varepsilon = 1 - \frac{\rho_M r + \rho_K c}{\rho_M x}, \quad (1)$$

where x is the penetration depth into a semi-infinite metal plate; r is the penetration depth into a ceramic-faced metal plate; c is the thickness of a ceramic plate; ρ_M is the density of a plate; ρ_K is the density of ceramics. The efficiency of construction (ε) is less then or equal to zero, if the mass of a composite plate does not exceed the mass of a target per unit area.

The efficiency of ceramics and composites was evaluated using the experimental data, numerical simulation [8-10] and formula (1) in the impact velocity range of 1...6 km/s, as shown in Fig. 4. The notations are used as follows: an averaged experimental dependence for Al_2O_3 ceramics at impact velocities of less than 2 km/s (1); ZrO_2 (2); corundum (3); Al_2O_3 (4); (TiB_2 + B_4C) (5); (TiC + NiCr)

(6), Al_2O_3 (AD85) (7). The experiments were carried out using the following projectiles: a 7.62 mm bullet (1); a ball (2-6); a rod (7). The results show that the maximum efficiency of ceramics corresponds to impact velocities up to 1.5 km/s. The efficiency of ceramics is decreased with increasing impact velocities. Some increase in the efficiency was observed for Al_2O_3 ceramics (corundum) and a (TiB_2 + B_4C) metal composite at impact velocities of 5-6 km/s. The main reason for the decrease in the efficiency of ceramics is the decrease in the strength of material due to the fracture of ceramics in shock wave.

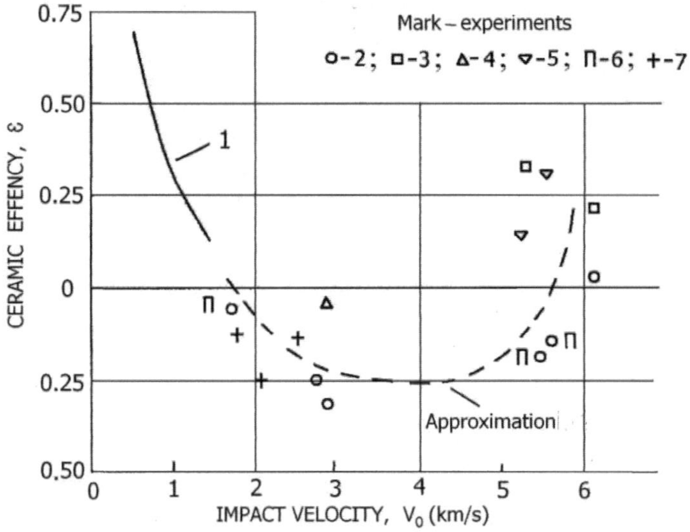

Figure 4. Efficiency of ceramics and composite materials at high-velocity impact

The penetration depth of a tungsten rod into layered targets was calculated considering weight-dimension characteristics. A target is a combination of a semi-infinite steel plate and ceramic plate with a thickness of c placed on the front surface of the main target (figure 3). A mass efficiency was calculated by the relation [13] according to the standards of ARL (Army Research Laboratory, USA):

$$e_m = \frac{(x-r)\rho_M}{c\rho_M} \tag{2}$$

The efficiency of ceramics δ with allowance for the weight-dimension characteristics of a layered structure was calculated by the imperial ratio:

$$\delta = e_m \frac{L_{cer}}{r}, \tag{3}$$

where L_{cer} is the penetration depth into homogeneous semi-infinite ceramics.

Figure 5 shows the calculation results demonstrating the efficiency of ceramics with allowance for the weight-dimension characteristics (δ) relative to steel. The efficiency values δ correspond to the ceramics as follows: AlN (1); Al_2O_3 (2); B_4C (3). Ceramics is considered to be effective if the value δ > 1. These dependencies clearly illustrate the high efficiency of composite materials in combination with steel plates against high-velocity rod projectiles at impact velocities up to 1.5 km/s. The efficiency of ceramics depends on mechanical and strength properties of material under dynamic loading. For a example, boron carbide is most effective among other ceramics investigated and reaches maximum impact velocities of 2 km/s.

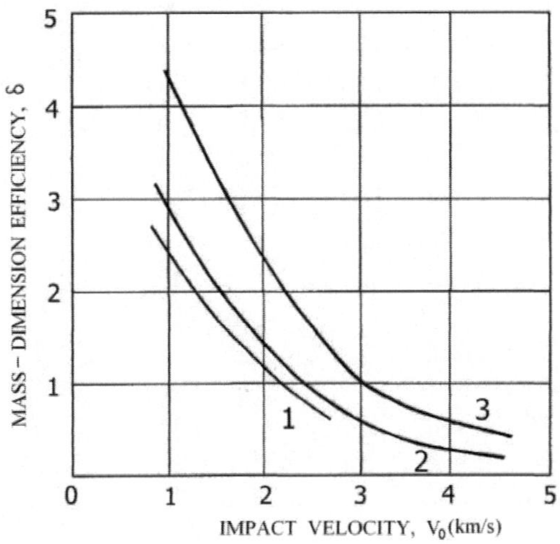

Figure 5. Weight-dimension characteristics of layered "steel-ceramics" targets at high-velocity impact

3. Conclusion

Thus, the authors of the paper developed the engineering methods evaluating the penetration of high-density rod projectiles into the elements of layered targets. The efficiency of ceramics such as boron carbide, aluminum oxide, and aluminum nickelide placed on the front surface of steel plates was evaluated in the wide range of high-velocity impact. The results have shown that the efficiency of ceramics depends on the mechanical properties of material under dynamic loading. The determining factors were revealed to explain the fracture and fragmentation of material in the contact area. Boron carbide was shown to be more effective ceramics compared to the other ceramics investigated and reached a maximum impact velocity of about 2.0 km/s.

The experimental results demonstrate the reasonable use of complex combined protection against high-velocity rod projectiles and the prospect for the strengthening of protection through the application of high-strength composite materials.

References

[1] Viechnicki D J, Slavin M J and Kliman M I 1991 *Ceramic bulletin* **70** 1035-1039
[2] Orphal D L, Frazen R R, Piekutowski A J and Forrestal V J 1996 *Int. J. Impact Engng.* **18** 335-368
[3] Subramanian R and Bless S J 1995 *Int. J. Impact Engng.* **17** 807-816
[4] Konjaev A A, Nazarov A G, Platova T M and Tolkachev V F 2000 *Proc. Conf. "Fundamental and Applied Problems of Modern Mechanics"* (Tomsk: Tomsk University Press) 83-84
[5] Konyaev A A and Tolkachev V F 2004 *Journal on Composite Mechanics and Design* **10** (4) 466-476
[6] Gorelsry V A, Zelepugin S A and Tolkachev V F 2000 *Chemical physics reports* **18** 2211-2217
[7] Tolkachev V F, Konyaev A A and Zheikov V V 2012 *Composites: Mechanics. Computation, Application. An International Journal* **3** 253-262
[8] Tolkachev V F 2016 High-velocity impact. Modeling and experiment ed. A V Gerasimov (Tomsk: Publishing house NTL) chapter 2 pp 57-112
[9] Tolkachev V F 2004 *Proc. Conf. "Fundamental and Applied Problems of Modern Mechanics"* (Tomsk: Tomsk University Press) 83-84
[10] Zelepugin S A, Tolkachev V F and Shpakov S S 2010 *Proc. Conf. "Modern ballistics and adjacent questions of mechanics"* (Tomsk: Tomsk University Press) 217-218
[11] Orlov M Y, Orlova Yu N and Tolkachev V F 2015 *J. Phys.: Conf. Ser.* **653** 012038

[12] Bless S J, Rosenberg Z and Yoon S 1987 *J. Impact Engng.* **5** 165-171
[13] Walters W, Gooch W and Burkins M 2000 Combustion , Explosion and Shock Waves **36** 745-750

IOP Conf. Series: Journal of Physics: Conf. Series **919** (2017) 012008 doi:10.1088/1742-6596/919/1/012008

Mathematical model and software for investigation of internal ballistic processes in high-speed projectile installations

A S Diachkovskii, A I Zykova, A N Ishchenko, V Z Kasimov, K S Rogaev, A D Sidorov

National Research Tomsk State University, 36, Lenin Ave., Tomsk, 634050 Russia

E-mail: Arven2022@mail.ru

Abstract. This paper describes a software package that allows to explore the interior ballistics processes occurring in a shot scheme with bulk charges using propellant pasty substances at various loading schemes, etc. As a mathematical model, a model of a polydisperse mixture of non-deformable particles and a carrier gas phase is used in the quasi-one-dimensional approximation. Writing the equations of the mathematical model allows to use it to describe a broad class of interior ballistics processes. Features of the using approach are illustrated by calculating the ignition period for the charge of tubular propellant.

1. Introduction

Applying of new high-energy propellants, complex designs of missile assemblies, ballistic installations on new physical principles requires constant improvement of mathematical models describing internal ballistic processes [1-3]. Mathematical models that adequately reflect the processes that occur during the shot process are most in demand in the formation of shot parameters predictive estimates in the extreme ranges of experiments.

In the presented mathematical problem formulation, a model of a polydisperse blend of non-deformable particles and a carrier gas phase moving along a channel of a spatial coordinate alternating cross section in a quasi-one-dimensional approximation is used [1]. During writing of the equations, it is assumed that the viscosity and heat conductivity of the gas are manifested only in interaction with the particles. It is also assumed that layer-by-layer burning of the powder propellant charge elements occurs at a velocity depended only on the temperature of the powder gases.

2. The basic system of gasdynamic equations

The basic system of equations that determines internal ballistic processes is written in a form that allows them to be used for a wide class of problems describing internal ballistic processes.

2.1. Equations expressing the laws of conservation of mass, momentum, and energy

It's as if there are particles of J varieties that differ in size and other characteristics. The gas is a homogeneous blend of J non-viscous non-heat-conducting gases with known equations of state, each of which is formed by the combustion of a particle of the corresponding kind.

Let us introduce the following notation:

t - time;

x - spatial coordinate measured along the axis of the axisymmetric barrel bore in the direction of throwing;

s(x)- channel cross-sectional area;

ρ_j – average density of the gases formed during the combustion of the particles j-th kind;

ρ_j^0 - истинная плотность *j*-го газа;

$\rho = \sum_{j=1}^{J} \rho_j$ - average density of a homogeneous blend of gases;

ρ^0 - true gas blend density;

$c_j = \rho_j / \rho$ - a mass fraction of the j-th gas in the gas blend $\left(\sum_{j=1}^{J} c_j = 1 \right)$;

α_j - volume fraction of j-th gas in the gas blend $\left(\sum_{j=1}^{J} \alpha_j = \alpha \right)$;

δ_j - true density of the j-th kind of particles;

β_j - volume fraction of the j-th kind of particles in the gas-powder blend $\left(\sum_{j=1}^{J} \beta_j = \beta = 1 - \alpha \right)$;

u_j - velocity of the j-th kind particles;

p, T, u - respectively, the pressure, temperature and velocity of the gas blend.

The motion of the polydisperse blend along the channel of the alternating cross section is described by the laws of conservation of mass, momentum, and energy, the form of which is given below.

For each component of the homogeneous gas blend

$$\frac{\partial}{\partial t}(\rho_j s) + \frac{\partial}{\partial x}(\rho_j u s) = m_{gj} s \qquad (j = 1, \ldots, J), \qquad (1)$$

summing which, we have the law of conservation of the gas phase mass

$$\frac{\partial}{\partial t}(\rho s) + \frac{\partial}{\partial x}(\rho u s) = \sum_{j=1}^{J} m_{gj} s \qquad (2)$$

For each kind of particles, we have

$$\frac{\partial}{\partial t}(\beta_j \delta_j s) + \frac{\partial}{\partial x}(\beta_j \delta_j u_j s) = m_{pj} s \qquad (j = 1, \ldots, J) \qquad (3)$$

There is $m_{gj}(m_{pj})$ - mass inflow of the j-th gas (particles) per unit time, per unit volume of the blend.

An equation

$$\frac{\partial}{\partial t}(\rho u s) + \frac{\partial}{\partial x}(\rho u^2 s) = -\alpha s \frac{\partial p}{\partial x} + 2\pi R \sigma_w^{n\tau} - s \sum_{j=1}^{J} (f_j - m_{gj} u_{gj}) \qquad (4)$$

is the law of conservation of momentum for the gas phase, and

$$\frac{\partial}{\partial t}(\beta_j \delta_j u_j s) + \frac{\partial}{\partial x}(\beta_j \delta_j u_j^2 s) = -s \beta_j \frac{\partial p}{\partial x} + s(f_j + m_{pj} u_{pj}) \qquad (5)$$

for the j-th fraction of particles.

Here the following notation is made: f_j - the force projection on the x-axis per unit blend volume with which the gas acts on particles of the first kind due to the difference in the phase velocities; $u_{gj}(u_{pj})$- velocity of the gas (particles) entering in the polydisperse mixture; $\sigma_w^{n\tau}$ - frictional stress in the gas at the channel surface.

For the gas phase, the energy conservation law in the form for the total energy has the form

$$\frac{\partial}{\partial t}(\rho E s)+\frac{\partial}{\partial x}(\rho u E s)=-\frac{\partial}{\partial x}(\alpha p u s)-\sum_{j=1}^{J}\frac{\partial}{\partial x}\left(p\beta_j u_j s\right)+$$
$$s\sum_{j=1}^{J}u_j\left(\beta_j\frac{\partial p}{\partial x}-f_j\right)+s\sum_{j=1}^{J}m_{gj}\left[Q_{gj}+\frac{u_{gj}^2}{2}\right]-sQ_r-sq \quad (6)$$

Here $E=e+\dfrac{u^2}{2}$ is total energy of gas blend;

$e=\sum_{j=1}^{J}c_j e_{gj}$ - internal energy of gas blend;

Q_{gj} – a specific internal energy of the gaseous combustion products entering the blend;

Q_r - an intensity of heat losses due to heat exchange with the walls of the barrel channel per unit volume of the blend;

$q=\sum_{j=1}^{J}q_j$, q_j – intensity of energy inflow to particles of the n-th grade per unit volume of the blend due to heat exchange with gas.

Similarly, for the j-th particles fraction

$$\frac{\partial}{\partial t}\left(\beta_j\delta_j E_j s\right)+\frac{\partial}{\partial x}\left(\beta_j\delta_j u_j E_j s\right)=-u_j\left(\beta_j\frac{\partial p}{\partial x}-f_j\right)+sm_{pj}\left[Q_{pj}+\frac{u_{pj}^2}{2}\right]+sq_j \quad (7)$$

where is $E_j=e_j+\dfrac{u_j^2}{2}$,

e_j – the specific internal energy of the particles (the known function of temperature);

Q_{pj} – the specific internal energy of the particles entering into the blend.

As can be seen from the structure of the equations, for the zero value m_{gj}, or m_{gj}, the corresponding values u_{pj}, u_{gj} and Q_{pj}, Q_{gj} can have any value except ∞, since they enter the equations as a product with m_{pj}, m_{gj}.

Equations (1) - (7) for given values m_{pj}, m_{gj}, u_{pj}, u_{gj}, Q_{pj}, Q_{gj} , f_j, q_j, which determine the interphase interaction of gas-condensed particles; $\sigma_w^{n\tau}$, Q_r - specifying friction and heat exchange on the surface of the barrel; Equations of state of the gas blend $e_j(T)$ and represents a closed system for determining the average densities of the gas components ρ_j, the velocity u_j, the volume content β_j and temperature of the particles T_j, the pressure p and the velocity u of the gas blend. The arbitrary choice of the dependencies that determine the interfacial interaction, friction and heat transfer to the barrel, the state equation of the gas components and the solid phase makes it possible to use system (1) - (7) for modeling a wide class of internal ballistic phenomena. A number of types of particles can be distinguished a priori, for which either these dependences are trivial, or some of the equations (1) - (7) do not make sense.

3. Classification of particles by types

3.1. Burning particle

Particles of this type are used to simulate the burning elements of a propellant charge, the corresponding quantities m_{gj}, ρ_j, ρ_j^0 are nonzero. For calculating combustion, the ratios generally accepted in internal ballistics are used [4]:

$$m_{pj}=-\frac{\beta_j\delta_j}{1-\psi_j}\frac{s_{0j}}{W_{0j}}\sigma_j(\psi_j)U_j(p)\,, \quad (8)$$

where ψ is a degree of transformation;

s_0 – initial particle surface;
W_0 – initial volume of a particle;
$\sigma(\psi)$ – ratio of the current burning surface to the initial;
$U(p)$ – law of layer-by-layer combustion.

Calculation of the energy equation (7) for a particle is not made $q_j = 0$.

3.2. Non-burning particle

For non-burning, the intensity of the volumetric mass input of the gas m_{gj} = 0, the corresponding values of the average ρ_j and true ρ_j^0 density of the gas formed during its burning are also zero, equation (1) has no meaning. This type of particle is used to simulate soot in combustion products, condensed ablation products, etc. In the presence of particles of this type, it is necessary to calculate the energy equation (7) for a particle, and to take into account the nonzero value q_j in the energy equation (6) for a gas.

3.3. Particle absent

This is the case of a fictitious particle that is introduced in order to uniformly calculate the gas originally located in the channel, for example, the products of the igniter's combustion in the event that its formation is not calculated. Equations (3), (5), (7) are not calculated, m_{gj} = 0.

The writing of equations in the form (1) - (7) allows one to have particles of any types in one computational domain. In this case, for example, a type 1 particle can form gaseous products and condensed components as a result of combustion, each of which in turn can be of type 2 (soot) or type 1 (TC dispersing products). Therefore, from the point of view of logical organization and computer realization it is advisable to introduce the term "particle sets" (PS), which means a certain closed set of particles. By closedness we mean that particles from a particular PS are, firstly, intended to describe a single element of a mathematical model, and second, they occupy the same continuous sequence of indices in an array of particles, and, thirdly, for them in the calculation process the quantities m_{pj}, m_{gj}, u_{pj}, u_{gj} are mutually consistent.

The simplest example of the PS is the model of the powder charge element, in the combustion of which only the gaseous component is formed. In this case, only one index $j = k$ corresponds to this "particles sets", the velocities of the burning particle and its combustion products equal $\left(u_{pk} = 0,\ \ u_{gk} = u_k\right)$, and the intensities of the bulk gas arrival and the decrease in the mass of particles are equal in magnitude $\left(m_{gk} = -m_{pk}\right)$ and are determined from (8).

In a more complicated case of combustion of a powder charge element with the formation of combustion products and one condensed component with a mass fraction, two components of a polydisperse blend (for example, one with an index $j = k$, the other with $j = l$) are required to describe this PS as the first for describing the element of the powder charge itself, for the resultant condensed particle. Here the following relations are satisfied

$$m_{gk} = -m_{pk},\quad u_{pk} = 0,\quad u_{gk} = u_k,$$

$$m_{pl} = -m_{pk}c,\quad u_{pl} = u_k,\quad u_{gl} = 0,\quad m_{gl} = 0,$$

where m_{pk} determined by (8).

Similarly, other "particle sets" are constructed, for example, to describe particles that are carried by a gas stream from a melting surface of a barrel or from dispersed products of TC.

The dependences that determine the strength and thermal interfacial interaction of gas and spherical particles are given in [1]. Introducing the particle bareness as a ratio of its surface to volume and writing the expression for f_j in a form generally accepted for the problem of external flow past a spherical particle, we have

$$f_j = \rho^0 \beta_j s_{pj} C_d \frac{|u - u_j| \, (u - u_j)}{8} \quad (9)$$

For grains of powders, taking into account the disorder in their packing and neglecting the influence of the velocity field of the combustion products flowing from the particle, a dependence of the form (9) is also used for the interphase interaction force.

For the resistance force of tubular and rod powders, a dependence of the form (9) is used, in which the resistance coefficient corresponds to the experimental dependences for packets of pipes that are streamlined in the longitudinal direction [5].

For the dependence of the particle surface on the transformation degree $\frac{s_0}{W_0}\sigma(\psi)$, a number of constraints are used that express the various geometric and physical laws of gas formation. A power pressure dependence is used for the linear combustion rate.

Friction and heat exchange with the barrel during gas motion are determined by semiempirical criterial dependencies obtained for the stationary flow regime. In this case, the temperature distribution in the thin heated layer of the barrel material is determined from the solution of the one-dimensional heat conduction problem in the cross sections of the barrel.

The system of equations (1) - (7) is solved numerically with the corresponding initial and boundary conditions.

Initial conditions:

$$p(x,0) = p_o; \qquad u(x,0) = 0; \qquad T(x,0) = T_o;$$

$$\rho_j(x,0) = \rho_{j0}; \quad (j = 0 \div J);$$

$$\beta_j(x,0) = \beta_{j0}(x); \quad u_j(x,0) = 0; \quad (j = 1 \div J),$$

where ρ_{0j}, p_0, T_0 are parameters characterizing the initial state of the gas phase;

$\beta_{j0}(x)$ sets the initial spatial distribution of the j-th fraction due to initial placement propellant elements.

In the simulation of internal ballistic processes, calculations performed on the developed mathematical model are carried out in the region bounded to the left by the charging chamber bottom cross section, and to the right by the projectile left end cross-section for the classical charging scheme or the combustion front for the charging scheme with the traveling charge.

In the section of the loading chamber bottom, the non-flow conditions are set.

When specifying the boundary conditions for the classical loading scheme, the projectile also specifies non-flow conditions. For the charging scheme with the traveling charge, the conditions given in [1] are used.

The above mathematical formulation is solved numerically by modifying the Godunov method, allowed to increase the approximation order on the spatial coordinate to the second order [6] and the time second order [7], and implemented as a software package (PC). When it was created, modern technologies for developing software systems were used.

Progress in the field of programming languages (PL) and development tools has made it possible to significantly simplify the development process and give the software new properties. At present, a more modern object-oriented approach has come to replace the technology of structural programming implemented in the PL, such as Algol, Fortran, Pascal, which facilitates programming, debugging and testing, support and further modification. The most convenient for implementing software systems developed within this approach are the PL, which support the style of object-oriented programming. Among them, the most common is C ++, which has a number of advantages over other object-oriented languages. Firstly, it is a compiled PL that combines the features of a high-level PL with the ability to perform low-level data operations and allows you to create high-speed programs. Secondly, compilers with C ++ exist on almost all computing platforms.

IOP Conf. Series: Journal of Physics: Conf. Series **919** (2017) 012008 doi:10.1088/1742-6596/919/1/012008

The essence of the object-oriented approach in the development of software systems for modeling the concepts and phenomena of the surrounding world consists in constructing their abstract models based on the selection of properties essential for the development purpose. If the simulated phenomenon is complex, then the models of its components are built, and the developed software system is considered as a composition of abstract models of its components. In C ++, the representation of an abstract model is a class, in fact it is a data type defined by a developer programmer to model a concept or phenomenon. The class contains data components and methods (functions) for working with data. An instance (an *object* in C ++ terms) of a class can be created in the main memory of a computer. Classes can be among themselves in relation to generalization or *inheritance*, allowing to identify analogies between them and to build a multi-level classification in relation to "general - private" (*class hierarchy*). A class that is closer to the top of the hierarchy (base class) is the bearer of more general properties with respect to classes located further from the top of the hierarchy (derived class). This is expressed in the fact that an object of a derived type always contains a subobject of the base type in its composition, while the same properties (data-components) for objects of the base and derived types have the same offsets relative to the initial address of the object. This allows you to view derived type objects from general positions (as objects of the base type), while using the technology of *virtual functions* when working with objects allows you not to lose the individuality inherent in objects of the derived type. It is possible to associate types belonging to different hierarchies for collaboration, using pointers (that is, addresses) to these or basic types. The set of hierarchies of classes constructed in this way, characterizing various aspects of the modeled domain, determines the functional content of the software package (PS). Obviously, using this development technology makes it easy to modify the composition of software by adding or excluding classes from the hierarchy. Very important is the ability to construct and destroy objects of an abstract type dynamically, while the program complex is running, which allows you to create a set of objects for solving a particular problem, guided by the specification of a set of input data.

With respect to mathematical modeling of processes in gas-dynamic propellant devices, the first level of abstraction is determined by the choice of physico-mathematical models describing the phenomenon and numerical methods for their solution. The described technology of software systems development is applied at the stage of software implementation for computer modeling of the subject domain under consideration. To do this, you need to define, for example, the following hierarchy of concepts, taking into account the multivariate types involved in the hierarchy:

- the calculation area together with the system of calculated equations and the calculation method (abbreviated as "calculation area");
- boundary condition (abbreviated as "boundary");
- installation geometry;
- resistance law;
- laminar burning rate;
- relative particle area;
- particle;
- multitude of particles;
- computing task;
- result output.

When implementing the software, the Qt library [8] was used, this library allows to create mobile software for various operating systems based on a single code written in PL C ++.

The software package is an application that runs on Windows. The main window of the PC can be in one of two modes: in the activated mode of calculation (in this case, the graphic output, which illustrates the calculation process, outputs to the client part of the main window) and in the inactive mode. In the first case, within the computational process, a workflow is created that performs calculations, the primary stream serves the output of graphic information and the operation of the data preparation subsystem (DPS). Thus, the software package is a two-stream application for Windows.

During the calculation, the execution of both threads is synchronized using standard Qt (mutexes and waiting conditions). The toolbar view in the calculation execution mode is complemented by the workflow management tools and visualization of the calculation results. In the inactive mode, the following options are available on the toolbar: calling the settings window and the DPS window.

The subsystem of data preparation allows to work on creation, editing and saving of a set of computational configurations, which represent a data set characterizing the realized mathematical and information model. The sets of data included in the design configuration must have the property of mutual consistency. This means that, being formally independent, they enter into a single design configuration, where different parts of the mathematical and information model implemented in the PC cannot be set independently of each other. An example is the placement of elements of a propellant charge whose left boundary cannot be less than the coordinate of the powder chamber bottom specified by the type "Internal geometry of the barrel", and the right boundary cannot be greater than the right boundary of the design area, which is determined by the types "Projectile with friction and backpressure "and "Travelling charge".

The difference between the developed mathematical model and those given in [2, 3] is that the description of internal ballistic processes is based on a single system of equations written for an arbitrary number of particles, each of which is included in the MF. The software implementation allows specifying, changing and adding to the program complex new possibilities practically without processing the existing program code. This can be illustrated with the example of calculating the pipe powder charge ignition period in the formulation [3].

The ignition period calculation was carried out on the basis of the mathematical model given in [3]. It is known that smoke gun powder (black powder) is often used as an igniter. The peculiarity of such powders is their rapid ignition, high combustion temperature with relatively low caloric content, as well as the content in the combustion products of a significant amount of solid particles K2SO4, K2S, K2SO3, C, etc. Moreover, the condensed phase is 50-60% of the total mass of the combustion products of black powder. The presence of heated particles leads to the appearance of significant heat fluxes to the powder charge surface during their depositions, which are commensurable with convective heat fluxes.

In terms of this mathematical formulation for describing black powder and its combustion products is used a combination of two components of a polydispersed blend of: the actual black powder particle with index n and formed as a result of its combustion soot index l. It is believed that the combustion of black powder occurs according to the law (8), and the rate of formation of soot particles per unit volume and their mass velocity are determined by the relations

$$m_{pl} = m_{pn}\left(1 - c_{gl}\right) - m_c, \quad u_{pl} = u_n,$$

where c_g is mass fraction of the gas formed during combustion of the black powder particle.

Here m_c is a rate of soot deposition on the surface of the ignited powder element per unit volume, connected with the analogous parameter G_c from [3] by the relation

$$m_c = \frac{G_c}{s}$$

The mathematical model used in comparison with [3] allows to take into account the polydisperse composition of the gas-powderblend, therefore the formulas from [3] should be generalized to this case. Passing from the perimeter of combustion to the bare spot s_{pj} of the powder element with the index j, this dependence can be represented in the form

$$G_{cj} = K_0 \frac{\delta_l \beta_l}{\alpha} s s_{pj} \beta_j \left|u_l - u_j\right|.$$

In this case, the total velocity of soot mass change due to precipitation in a unit volume will be

$$m_c = \sum_B G_{cj} = \frac{\delta_l \beta_l}{\alpha} \sum_B K_0 s_{pj} \beta_j \left|u_l - u_j\right|$$

Here the summation is carried out over all indices j ignited but not yet burning propellant elements. For the sedimentation coefficient K_0, the same functional dependence is used as in [3].

Assuming that the heat flow is due to convective heat transfer from the relative motion of the gas flow near the powder element and the conductive heat transfer due to the deposition of black powder particles, and adopting the principle of additivity of the heat flow components, we have an expression for the heat flux into the ignited kind j element

$$q_j = (\alpha_{cj} + \alpha)(T - T_{wj}),$$

where T is gas blend temperature;

T_{wj} – surface temperature of the ignited powder element.

For the convective heat transfer coefficient we have the known relation

$$\alpha = \frac{\lambda Nu_j}{d}.$$

Here λ is coefficient of gas thermal conductivity;

Nu_j – the Nusselt number, known flow conditions function;

d a characteristic linear dimension, which for a packet of tubes can be represented in the form

$$d = \frac{4\alpha}{\sum_B s_{pj}\beta_j}$$

The expression for the heat flux is used as the boundary condition in the tack of surface igniting of the powder element of grade j on the basis of a solid-phase ignition model. For each of the igniting, but have not yet burning powder particles to solve the equation

$$\frac{\partial T_j}{\partial t} = \frac{\lambda}{\delta c}\frac{\partial^2 T_j}{\partial n^2} + \frac{Q_c z}{c}\exp(-E_a/RT_j) \qquad (10)$$

where $T_j = T_j(x,n,t)$;

n – spatial coordinate directed along the normal to the particle interior;

c – specific heat capacity;

Q_c – thermal effect of the reaction in the condensed phase;

z – pre-exponential factor;

E_a – activation energy,

under the following initial and boundary conditions:

$$T_j(x, n, 0) = T_0; \quad n \geq 0;$$

$$-\lambda\frac{\partial T_j}{\partial n} = (\alpha_{cj} + \alpha)(T - T_j); \quad n = 0;$$

$$\frac{\partial T_j}{\partial n} = 0; \quad n \to \infty,$$

where T_0 is initial temperature of propellant.

Equation (10) is solved until the moment of ignition, in which quality the surface of the particle is used to reach a sufficiently high temperature T*. After that, the particle is considered to be burning and the gas flow from it is calculated by (8).

As an example, figures 1 and 2 show the results of calculations for two tube powders types as 22/1 and 11/1.

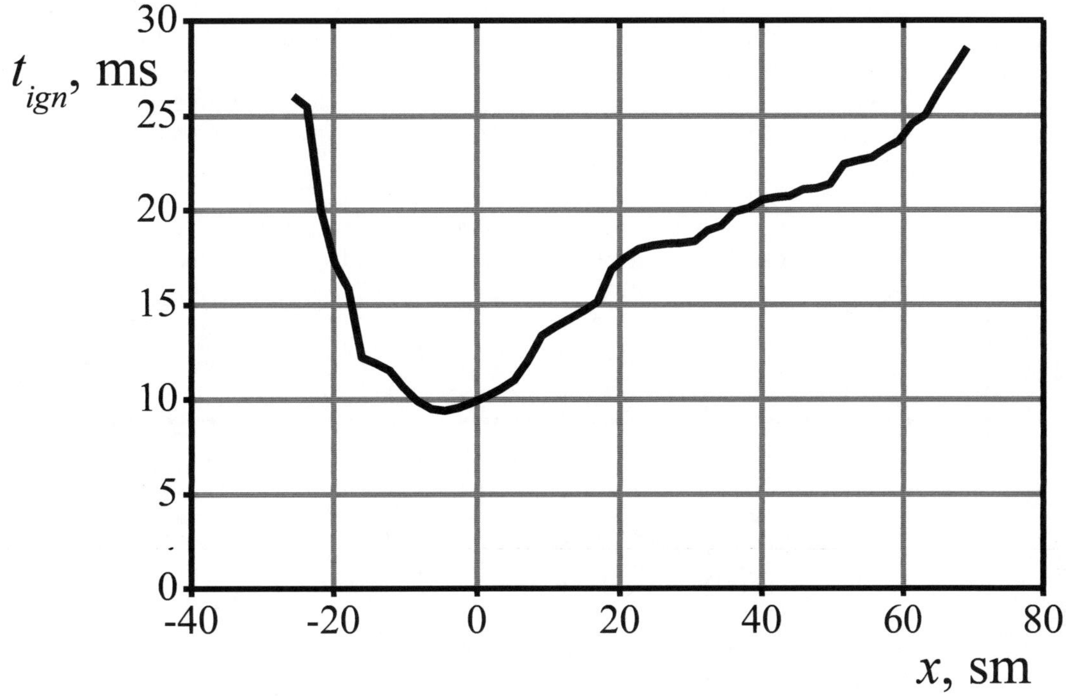

Figure 1. Charge ignition time for powder 22/1.

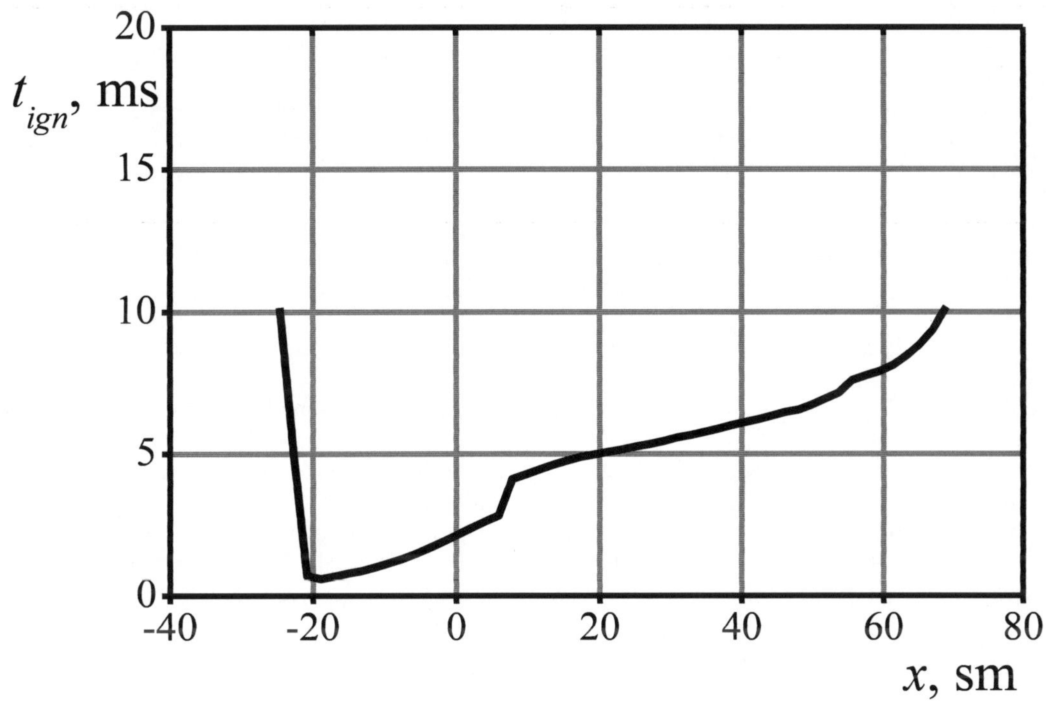

Figure 2. Charge ignition time for powder 11/1.

The times of charge elements ignition for different cross sections of the charge are shown. During these calculations, the black powder mass was significant relative to the main charge mass. As a result, the sections located at the charging chamber bottom are ignited later than in the middle of the chamber. This is explained by the increase of the heat flux in the chamber middle due to the convective component, which depends on the powder gases velocity. Powder with a smaller depth of vault (11/1) ignites faster than 22/1.

Figure 3 shows the results of calculations for a variant more appropriate to the real masses ratios of black powder and the main charge. One igniter was located at the chamber bottom, the second was in the middle of the main charge. A solid curve shows the dependence of the ignition time on the cross

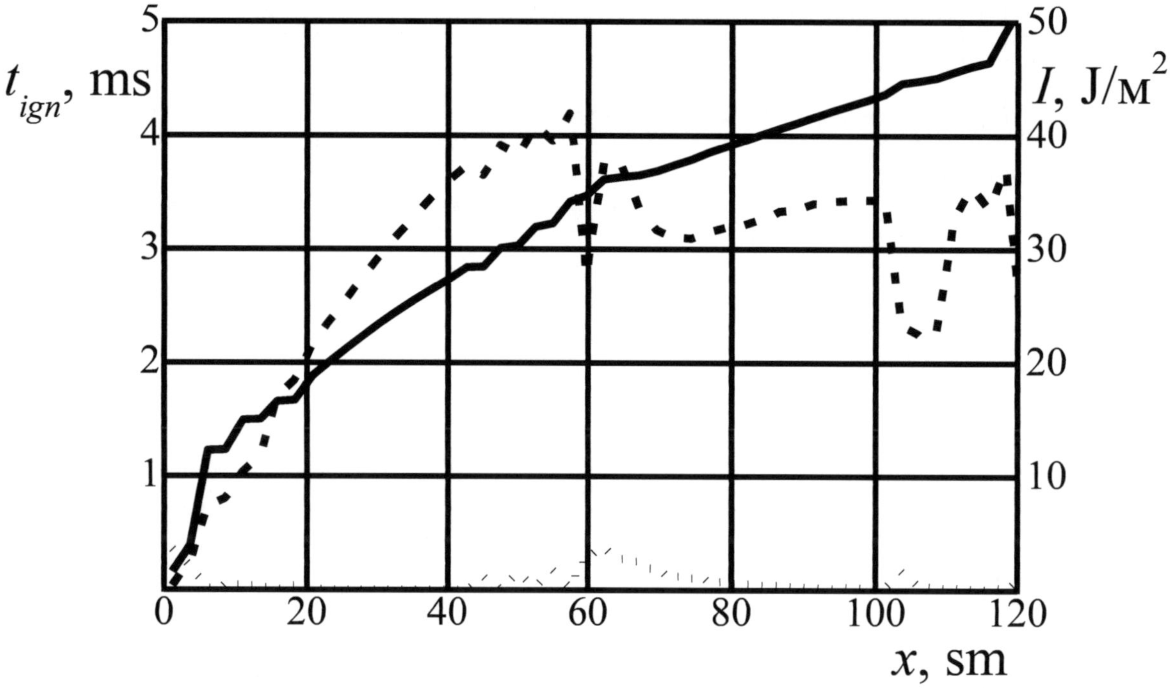

Figure 3. Ignition time of the charge and the energy per unit of the ignited surface for the conductive and convective warm-up mechanism

section coordinate. A dashed curves and dots represent respectively the values of the integrals $I_{convective} = \int_0^{t_{ign}} \alpha\left(T - T_{wj}\right)dt$ и $I_{conductive} = \int_0^{t_{ign}} \alpha_{cj}\left(T - T_{wj}\right)dt$, representing the amount of energy per unit of the surface to be ignited. In this case, the leading mechanism in the process of whole charge ignition is the conductive mechanism due to the deposition of hot particles of black powder. Its role is essential for igniting the part of the charge that is located at the bottom of the charging chamber. Here, the $I_{conductive}$ value significantly exceeds $I_{convective}$, which determines the short ignition time of the main charge elements located here. Subsequently, the powder gases flowing from the charging chamber bottom increase the convective heat flow and the role of the convective mechanism for igniting the main charge elements located at a distance from the charging chamber bottom is the leading one.

4. Conclusion

In this article, there was described the mathematical model intended to investigate the wide class of internal ballistic processes in high-speed propulsion systems. Its elements and general approach were used in modeling processes in light-gas installations [9]. The implementation in the form of the software package allowed it to be used by its personnel who are not specialists in the field of mathematical modeling.

Acknowledgement
The results were obtained within the framework of the state task of the Ministry of Education and Science of the Russian Federation, project No. 9.9036.2017/8.9.

References

[1] Khomenko Yu, Ishchenko A, Kasimov V 1999 *Mathematical modeling of internal ballistic processes in barrel systems* (Novosibirsk: Publishing house SB RAS) p 256
[2] Rusyak I and Ushakov V 2001 *Intracameral heterogeneous processes in barrel systems* (Ekaterinburg: URAN) p 259
[3] Rusyak I, Lipanov A, Ushakov V 2016 *Physical basis and gas dynamics of gunpowder combustion in artillery systems* (Moscow Izhevsk: Institute for Computer Research) p 456
[4] Serebryakov M 1962 *Internal ballistics of barrel systems and powder rockets* (Moscow: Oborongiz) p 703
[5] Idelchik I 1975 *Reference book on hydraulic resistance* (Moscow.: Mechanical Engineering) p 559
[6] Colgan V 1974 Application of the derivative minimum values principle to the construction of finite difference schemes for calculating discontinuous solutions of gas dynamics *Scholarly notes Central Institute of Aerohydrodynamics vol 3 No 6* pp 68-77
[7] Bogdanoff D and Miller R 1995 *New Higher-Order Godunov Code for Modelling Performance of Two-Stage Light Gas Guns* NASA TM 110363 pp 45
[8] Blanchet J and Summerfield M 2008 *Qt 4: Programming the GUI in C ++* (Moscow: KUDITS-PRESS) p 736
[9] Kasimov V, Ushakova O, Khomenko Yu 2003 Numerical simulation of intraballistic processes in a light-gas gun *Applied Mechanics and Technical Physics vol 44 No 5* pp 13-22

IOP Conf. Series: Journal of Physics: Conf. Series **919** (2017) 012009 doi:10.1088/1742-6596/919/1/012009

Determination of the propellant combustion law under ballistic experiment conditions

A N Ishchenko, A S Diachkovskii A I Zykova, VZ Kasimov, N M Samorokova

National Research Tomsk State University, Tomsk, Russia

E-mail: Arven2022@mail.ru

Abstract. The main characteristics of ballistic experiment are the maximum pressure in the combustion chamber $P_{\max}$ and the projectile velocity at the time of barrel leaving U_{M}. During the work the burning law of the new high-energy fuel was determined in a ballistic experiment. This burning law was used for a parametric study of depending $P_{\max}$ and U_{M} from a powder charge mass and a traveling charge at initial temperature of + 20 °C was carried out. The optimal conditions for loading were obtained for improving the muzzle velocity by 14.9 %. Under optimal loading, there is defined the conditions, which is possible to get the greatest value muzzle velocity projectile at pressures up to 600 MPa.

1. Introduction

In the internal ballistic design of artillery systems, various approaches to increasing the projectile velocity at the exit from the barrel are being investigated. One such approach is throwing a projectile using a traveling charge (TC). This work contributes to the improvement of traditional artillery weapons.

There are classical and non-traditional schemes of throwing. In the classical throwing scheme, the projectile acceleration occurs only due to the powder charge combustion in the chamber. One of the non-traditional schemes of throwing is a scheme with a TC. In this scheme, perspective high-energy convective combustion propellants are used as a traveling charge. The traveling charge is located directly behind the projectile and they move together along the barrel channel, forming a two-body system Projectile + TC. Figure 1 shows that, during motion along the cylindrical channel, the projectile is exposed by gases with gradual fuel combustion in two regions where the traditional powder charge and TC are located.

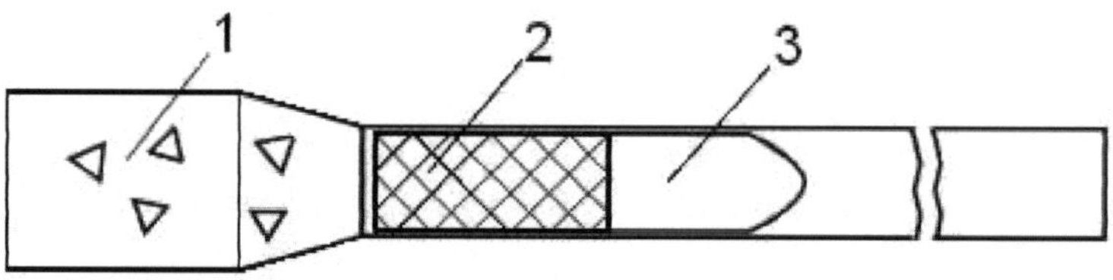

1 – Conventional powder charge,
2 – traveling charge, 3 – projectile,
Figure 1 – General scheme of throwing with TC

Initially, the main powder charge is ignited in the combustion chamber. The projectile starts to move, and after some time called the ignition delay time, the traveling charge ignites and burns from the end remote from the projectile. In the acceleration process the throwing gas pressure changes due to two main factors. The gases arrival due to propellant elements burning increases the pressure, and the increase of volume behind the projectile reduces it. Maintenance of pressure at a sufficiently high level is possible by using progressive combustion elements in the propellant charge. In this respect, new pasty progressive combustion character propellants have great prospects.

Thus, the use of TC allows to increase the muzzle projectile velocity without increasing the maximum chamber pressure. This is achieved by propellant charge energy increasing, increasing the pressure in the barrel depth directly in space behind the projectile and reactively projectile driving, which appears due to the TC combustion products outflow.

The purpose of this paper was to determine the combustion law of a promising high-energy propellant in a ballistic experiment at a standard laboratory temperature of + 20 ° C. Also there is a pursuance of the parametric research of the connection between loading parameters and ballistic parameters of the experiment. This study will assess the new propellant potential of projectile velocity increasing, as well as enter into work about a temperatures effect on the processes of throwing.

The main task of this study was to determine by how many percent the muzzle projectile velocity can be increased in a pressure range up to 600 MPa, permissible for the particular ballistic installation, and the necessary loading conditions.

1 - Combustion chamber,
2 - Breech support,
3 – Carriage,
Figure 2 – Ballistic installation

2. Experimental data processing

The main thermodynamic characteristics of propellants and combustion products: density, combustion temperature, force, covalent, adiabatic index and molecular weight are determined by the manufacturer. The law of propellant combustion under the closet vessel conditions is also defined.

However, for a more detailed study of propellant, there is a need to determine the propellants combustion law in the ballistic experiment, because combustion may be different.

Figure 2 shows a ballistic installation where laboratory experiments on classical and non-traditional throwing schemes were conducted during the work in the 70th department of the RIAMM TSU. A pressure in the combustion chamber P(t) and a maximum pressure in the chamber Pmax, a velocity of the projectile as it travels along the barrel channel U and at the moment of barrel departure UM were measured in experiments. The inert projectile used in the whole experiments series had a fixed mass. A granular powder charge was used as the main charge in the chamber. The TC made from the perspective paste propellant was placed in a polyethylene container. From one end the container is covered with a grid, which prevents the propellant from escaping. In experiments on the classical throwing scheme, a polyethylene imitator of the same mass and size as the TC in the container was used. Thus, the starting assembly mass, the initial projectile element position and, consequently, the pressure of forcing Pf (the pressure when the projectile was struck) was preserved in the all series of experiments.

Figure 3 shows the relationship between the combustion chamber pressure and the projectile as a function of time for different throwing schemes. As it can be seen the projectile velocities are practically the same at the initial stage. A point A is a moment of the rapid increase in the velocity, which corresponds to the onset of TC combustion.

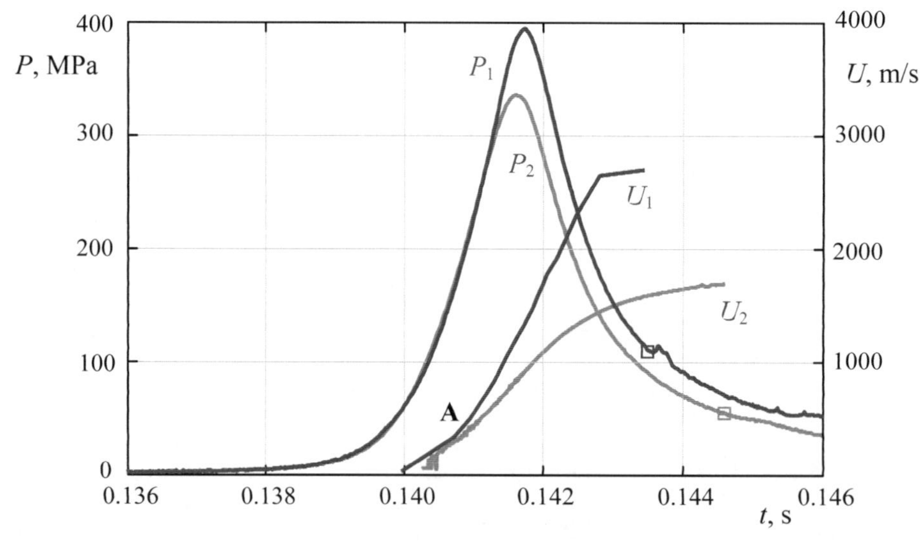

Figure 3. Projectile pressures and velocities dependences of time in experiments with the initial charge temperature +20°C
1 – non-traditional scheme with the TC, 2 – classical and scheme

A mathematical model developed in the 70th department of the RIAMM TSU was used to investigate the above throwing scheme. This model allows to solve the direct internal ballistics problem for various throwing schemes, to determine the projectile pressure and velocity for the specified parameters of the installation, charge and projectile. A special program complex [2], developed in the 70th department of the RIAMM TSU, was used for calculations.

At the first stage of the experimental data analysis, there was determined friction of barrel motion τ. Such parameters are determined as matching parameter of the experimental and calculated data based on experiments with the imitator. Id est during the series of calculations, this parameters are changed in certain range to achieve the maximum coincidence of the experimental (in Tables 1 and 2 are denoted by P_e, U_e) and the calculated values (denoted by P_c, U_c) of P_{max} and U_M, and respectively curves $P(t)$ and $U(t)$, as shown in Fig. 4. The standard gunpowder charge parameters are used.

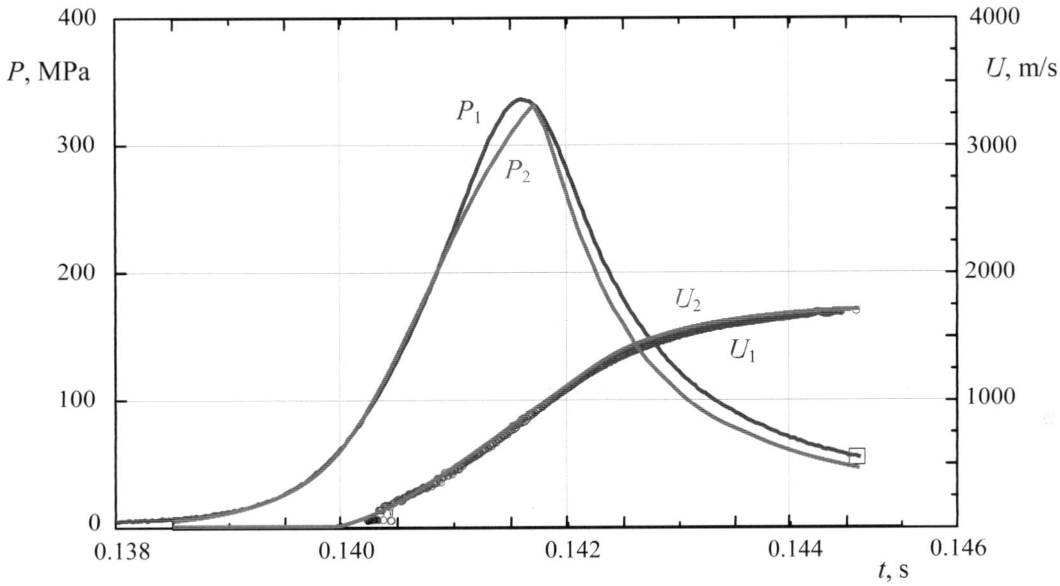

Figure 4. Typical plot of the chamber pressure (P) and the projectile velocity (U) versus time for the classical scheme of throwing
1 – experiment, 2 - calculation

Table 1 presents the matching parameters of the classical throwing scheme obtained in the calculation of the experiments at the standard temperature T_1 = 20 °C, and the difference between the calculated and experimental values of the maximum pressure (ΔP_{max}) and the muzzle velocity of the projectile (ΔU). Then the parameters were used in all calculations to determine the propellant combustion law used as the TC in an non-traditional throwing scheme.

Table 1. Matching parameters of the classical throwing scheme

P_f, MPa	τ, MPa	ΔP_{max}, %	ΔU, %
60	1	0.77	0.3

While the theoretical analyzing the gunpowder charge in the chamber is believed instantly ignites and burns at known linear velocity. The TC ignites after a certain time of gunpowder combustion. TC Combustion is determined by the impulse, which is calculated by the formula (1):

$$I = \int_0^t P dt \qquad (1)$$

were u – -combustion rate, B_i – velocity coefficient, P – pressure behind the combustion front, ν_i, – power factor.

an explosion point is set by the ignition delay impulse $I = I_1$. The TC combustion is characterized by two periods of slower combustion with a velocity $u = B_1 \cdot P^{\nu}$ and faster combustion with a velocity $u = B_2 \cdot P^{\nu}$. A transition moment is determined by reaching the pulse value of the accelerated combustion impulse $I = I_2$. The combustion law parameters as I_i, B_i depend on the fuel properties and they were determined as matching parameters of the experimental and calculated data (Fig. 5). Table 2 gives the obtained parameters of the combustion law.

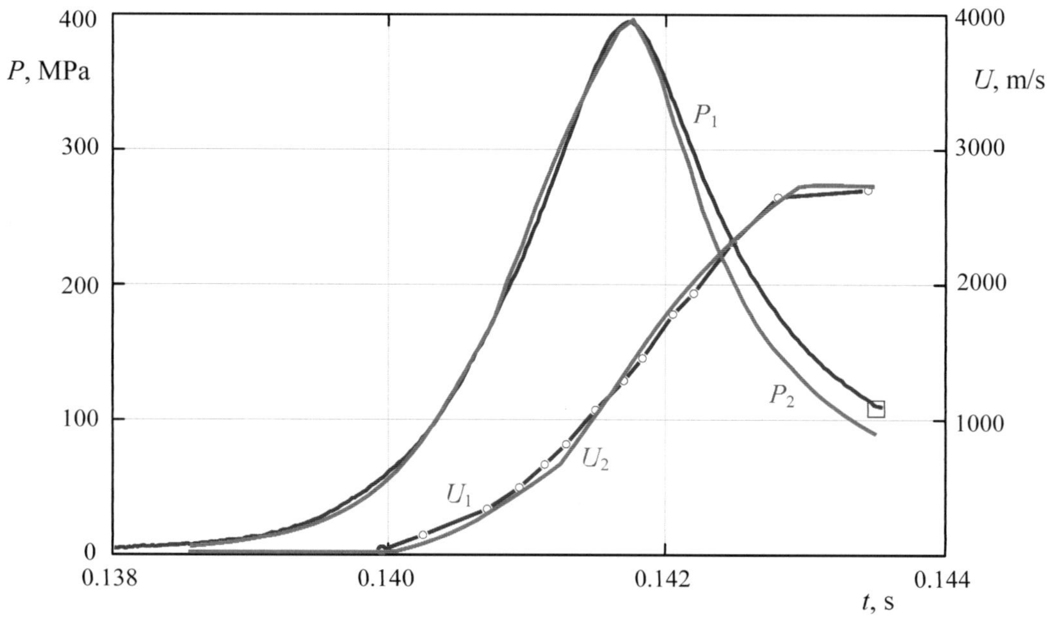

Figure 5. Typical plot of the chamber pressure (P) and the projectile velocity (U) versus time for the scheme with TC
1 – experiment, 2 - calculation

Table 2. Matching parameters for nontraditional scheme throwing with TC

B_1, m/s/(0.1·MPa)$^{\nu 1}$	I_1, MPa·s	B_2, m/s/(0.1·MPa)$^{\nu 2}$	I_2, MPa·s	ΔP_{max}, %	ΔU, %
0.0045	0.0865	0.024	0.19	0.7	0.3

The patterns of throwing obtained during the processing of experiments can be considered valid, since the discrepancy between the experimental and calculated values does not exceed 0.77%. These matching parameters were used to further investigation the relationship between the loading parameters and the muzzle velocity of the projectile.

3. Parametric investigation of the loading parameters

Using the obtained combustion law, parametric investigations were performed during which the mass of the powder charge in the chamber (Fig. 6) and the mass of the attached charge (Fig. 7) varied. The mass of the projectile and the TC remained unchanged. In the permissible pressure range (600 MPa), the velocity does not practically change, regardless of how much the mass of the powder or the TC increases. This is due to the TC combustion with such a velocity that the propellant doesn't have time to burn up before the projectile leaves the barrel.

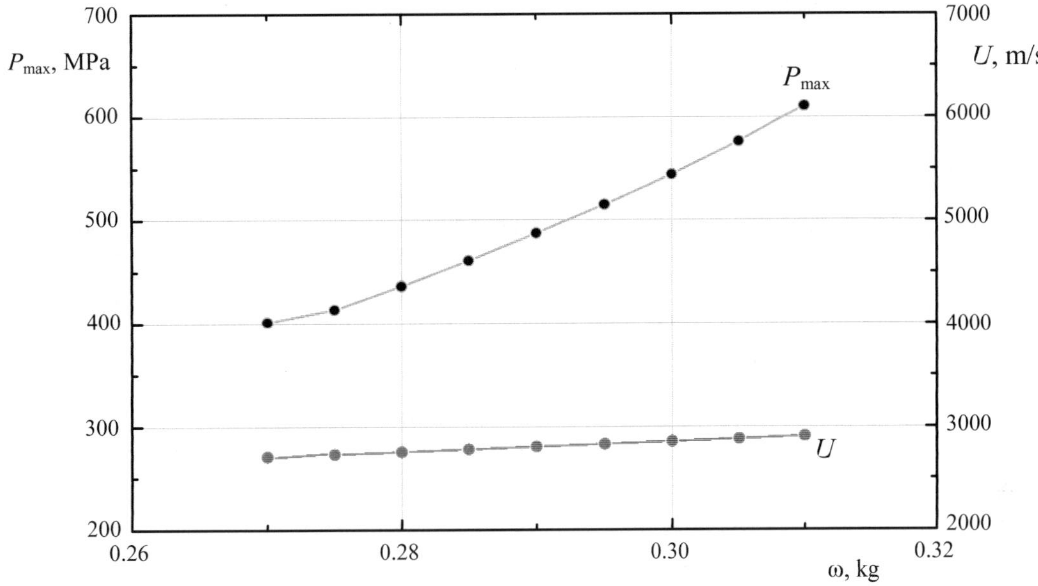

Figure 6. Dependence of the maximum pressure and muzzle velocity of the projectile on the change in the mass of the powder charge in the chamber

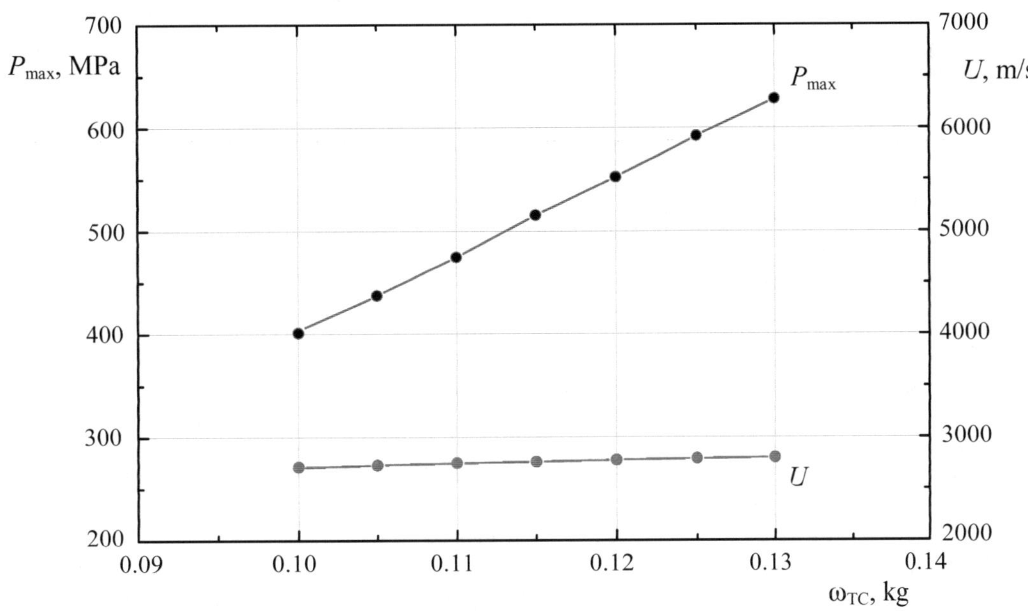

Figure 7. Dependence of the maximum pressure and muzzle velocity of the projectile on the change in the mass of the TC

In connection with the characteristics of propellant combustion, the optimum result by using the TC is obtained when the ignition of the TC occurs after the pressure peak. There are mechanical and chemical mechanisms for delay, but the most practical one should be considered the chemical retarder design. In the calculations, the presence of the chemical retarder is modeled by the accelerated combustion impulse I_2. In the course of the study, the delay impulses changed along with the mass of the gunpowder and TC, the variation in muzzle velocity and maximum pressure in the combustion chamber was checked.

Figure 8 shows the dependence of the maximum possible velocity on the specified maximum pressure in the chamber at T_0 = 20 °C. Thus, Figure 8 demonstrates the optimal accelerated combustion impulses I2 selected for the initial temperature of the charge + 20 °C. At these impulses the best initial velocity values in the admitted pressure range are reached. Line 1 is the results obtained by changing the mass of the gunpowder charge in the chamber at the constant mass of the TC, I_2 = 0.26 MPa·s. Straight lines 2 and 3 are the results obtained with the constant mass of the powder charge in the chamber and with TC mass changing, the accelerated combustion impulses I_2 are respectively 0.26 MPa·s and 0.3 MPa·s respectively.

It can be seen that the variation of the impulse I_2 makes it possible to increase the maximum possible velocity at the particular maximum pressure in the chamber.

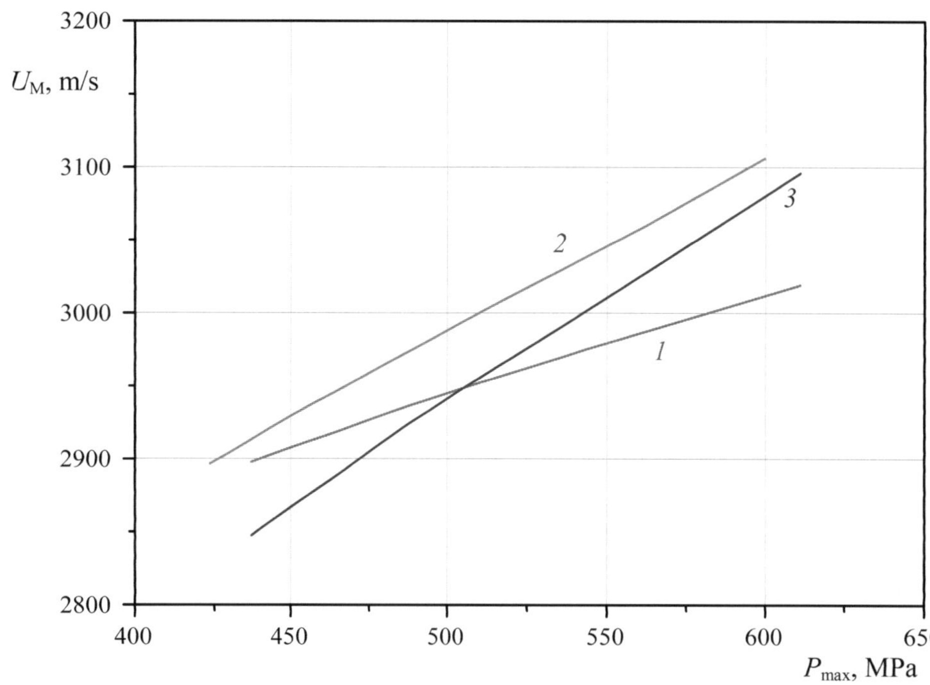

Figure 8. Dependence of the initial velocity on the maximum pressure at different impulses I2, gunpowder mass and TC mass (T_0 = + 20 °C)
1 – I2 = 0.26 MPa·s; 2 – I2 = 0.26 MPa·s ; 3 – I2 = 0.3 MPa·s

4. Investigation results

The complex study of propellant, as well as its potential capabilities under given laboratory conditions, was carried out

As a result of processing of the experimental data results obtained using the classical throwing scheme and the-throwing scheme with the TC at an initial charge of + 20 ° C, the matching parameters were obtained on the basis of a comparison of the experimental and calculated data, with a discrepancy of not more than 0.77%.

The parametric investigation was carried out to determine the possibility of increasing the projectile muzzle velocity in the pressure range not exceeding 600 MPa. Specific values of the variable parameters were obtained at which the muzzle velocity was increased by 14.9% relative to the value obtained in the experiment.

5. Conclusion

The results of the study were obtained for the standard laboratory temperature and were used later to investigate throwing at lower (-50 °C) and higher (+ 40 °C) temperatures to determine the temperature coefficients of the propellant. This made it possible to carry out a study on the possibilities of modernization of the throwing using this propellant for different temperatures.

Acknowledgement
This research was supported by grant (№ 8.2.05.2017) from "The Tomsk State University competitiveness improvement program.".

References

[1] Ishchenko A and Kasimov V 2015 *Mathematical model and program complex for theoretical research of internal ballistic processes in barrel systems* (Tomsk: Tomsk State University Publishing House) pp 21-27

[2] Kasimov V 2005 Program complex for calculation of internal ballistic processes in barrel systems *Journal of Russian Academy of Missile and Ammunition Sciences* №1 pp 70-76

[3] Serebryakov M 1939 *Internal ballistics* (Moscow: State publishing house of the defense industry)

[4] Rusyak I and Ushakov V 2001 *Intrachamber heterogeneous processes in barrel systems* (Ekaterinburg: Ural branch of the Russian Academy of Sciences publishers)

[5] Betekhtin S, Vinitsky A, Gorokhov M 1957 *Gasdynamic bases of internal ballistics* (Moscow: Oborongiz)

[6] Corner J 1950 *Theory of the interior ballistics of guns* (New York: Publishing House John Wiley & Sons, inc.)

[7] Ansari K, Baugh J Jr 1988 Dynamics of a Balloting Projected in a Moving Gun Tube *Ballistic research laboratory aberdeen proving ground* (Maryland)

[8] Brode H and Enstrom J 1970 *A numerical method for calculating interior ballistics* (USA California Santa Monica: The RAND Corporation)

[9] Gough P 1993 Formulation of next-generation interior ballistic code *Aberdeen proving ground report* (Marylan: Ballistic Research Laboratory)

[10] Jedlička L and Beer S 2007 Pressure distribution in the space behind a projected *Scientific article (Advances in military technology Issue 1)* pp 35–48

[11] Acharya R and Kuo K 2007 Finite element simulation of interior ballistic processes in 120-mm mortar system *Research paper (23-rd international symp. on ballistics)* (Tarragona)

[12] Baer P and May I 1988 Traveling-Charge Effect Gun Progress *Astronautics and Aeronautics Propulsion Technology vol 109)* (Washington: AIAA Inc.)

[13] Ermolaev B, Sulimov A, Roman'kov A *Traveling high-density charge of convective burning in the hybrid configuration: new results* (Moscow: Semenov Institute of Chemical Physics Russian Academy of Sciences)

IOP Conf. Series: Journal of Physics: Conf. Series **919** (2017) 012010 doi:10.1088/1742-6596/919/1/012010

Determination of extra trajectory parameters of projectile layout motion

A Ishchenko[1], V Burkin[1], V Faraponov[1], L Korolkov[1], E Maslov[1,2], A Diachkovskiy[1], A Chupashev[1], A Zykova [1]

[1] Tomsk State University, 36 Lenina Avenue, Tomsk 634050 Russian Federation
[2] Tomsk Polytechnic University, 36 Lenina Avenue, Tomsk 634050 Russian Federation

E-mail: maslov_eugene@mail.ru

Abstract. The paper presents a brief description of the experimental track developed and implemented on the base of the RIAMM TSU for external trajectory investigations on determining the main aeroballistic parameters of various shapes projectiles, in the wide velocity range. There is comparison between the experimentally obtained dependence of the fin-stabilized projectile mock-up aerodynamic drag coefficient on the Mach number with the 1958 aerodynamic drag law and aerodynamic tests of the same mock-up

1. Introduction

One of the main parameters that characterize the body motion in the air is an aerodynamic drag force coefficient (*Cx*) [1]. An experimental ballistic track for research aimed to identify aeroballistics parameters, including the *Cx* of different shapes projectile layouts, is developed and implemented in TSU Research Institute of Applied Mathematics and Mechanics.

2. Description of the experimentall installation and a technique of conducting experimental researches

A smoothbore ballistic installation is used as an accelerator (Fig. 1, 1). The accelerator has on the muzzle an induction sensor of projectile assembly muzzle velocity [2]. A barrel partly recessed into an evacuated suppressor (Fig.1, 2).

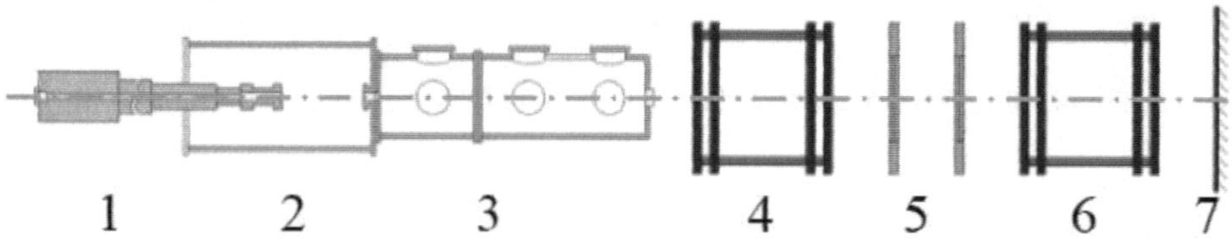

Figure 1. Schematic disposition of track basic elements.

The suppressor is matched with a windows section (Fig. 1, 3) for video and lighting division projectile assembly process. A master device fragments clipper is installed on the section output. There are two photoelectron velocimeters FEB-7M (Fig. 1, 4, 6) after clipper on the projectile axis at a certain distance from each other. Frames with paper target-marks are mounted on the area between the velocimeters on the guide fixture (Fig. 1, 5). These target-marks are serving as needed to monitor the projectile attitude on the trajectory. A bullet catcher shield is behind the second FEB-7M (Fig. 1, 7). The total length of the element flight path is 27 m.

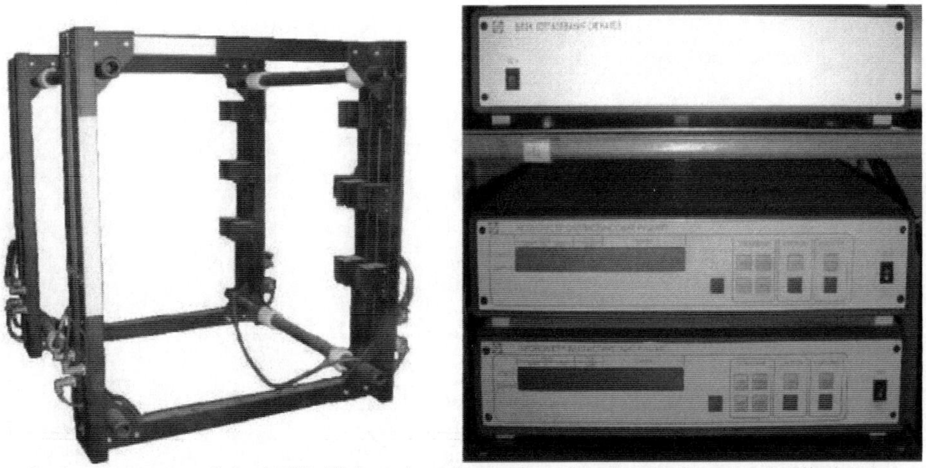

Figure 2. Appearance of the FEB-7M, a signal synchronization unit and a ballistic chronometers "RUSH-MP".

Developing ballistic track hardware allows: to detect the accelerator chamber pressure by a piezoelectric pressure sensor; to determine the projectile assembly velocity at the exit of the accelerator by the inductive muzzle velocity sensor; to produce a high-speed video of the processes accompanying the projectile motion on the trajectory; to measure the projectile velocity on the two sections of the trajectory by the velocimeters FEB-7M. In the open part of the trajectory, projectile velocity is registered by microwave radar.

Figure 2 shows the appearance of the FEB-7M, a signal synchronization unit and a ballistic chronometers "RUSH-MP".

Used photoelectron velocimeter allows to determine the projectile velocity in the range of 100 ... 2000 m/s at fractional accuracy less than 0.15%.

Figure 3 shows a typical finned projectile layout tested on the ballistic track. The element is made of duralumin with screw weighted steel head.

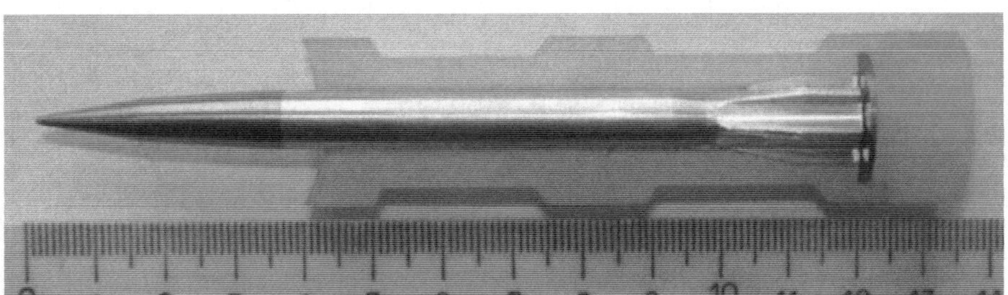

Figure 3. Appearance of the FEB-7M, a signal synchronization unit and a ballistic chronometers "RUSH-MP".

3. Results of experimental studies

Before the test the layout was placed in the polyethylene master device made for it, which consisting of a pair of beams, steel labels required for the flight time registration by the induction muzzle velocity sensor assembly and a pushed plastic pallet.

Moment of master devices disclosure to clipper was fixed speed video. Fig. 4 shows the projectile layout moves without apparent attack angle, and master device fragments were correctly separated without introducing further movement perturbations.

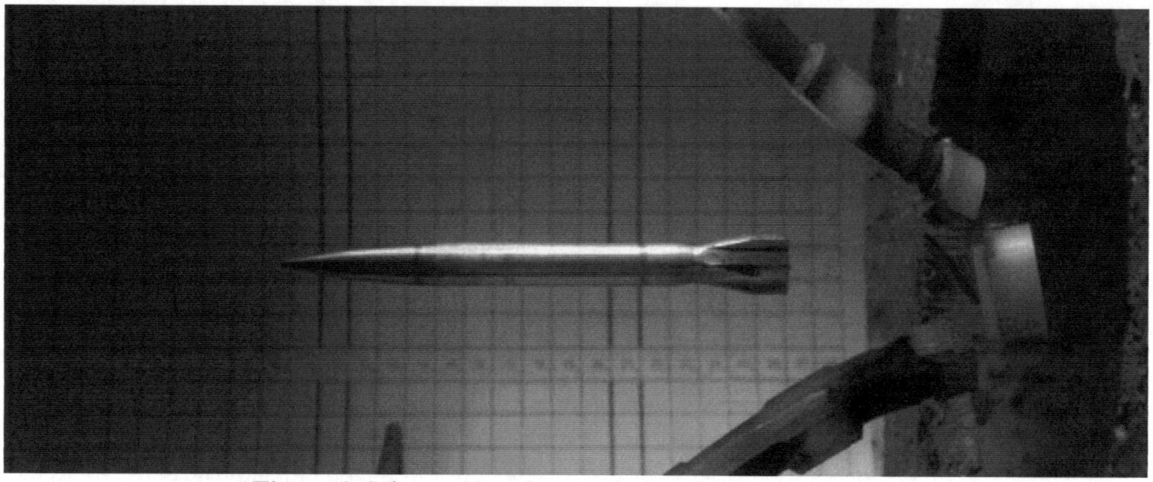

Figure 4. Scheme (a) and general view (b) of cone model.

There is a series of 14 tests to determine the aerodynamic drag force coefficient C_x of the similar projectile layout. The same way during the tests temperature and atmospheric pressure was recorded in the ballistic track room.

Determination of the C_x is conducted in accordance with the scheme Figure 5.

Before the performance of the tests series, the distance between the centers of photoelectric velocimeteres was measured by a laser ranger. The obtained value was used in the calculations as the distance L between the velocity measuring points V_1 and V_2.

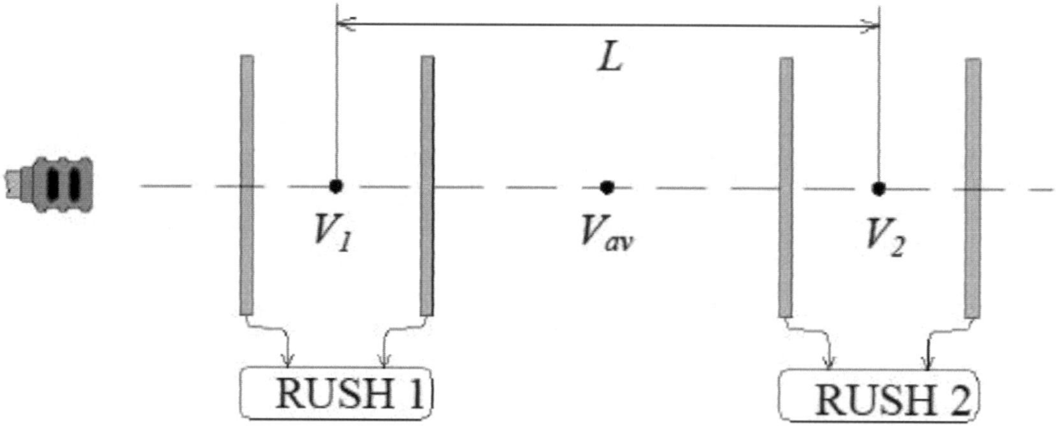

Figure 5. Scheme of the ballistic track to calculate the value of C_x.

The average value of the coefficient of layout aerodynamic drag forces C_x on the considered distance L is determined by the relation [3]:

$$C_{Xav} = \frac{2m}{\rho SL} \ln \frac{V_1}{V_2},$$

where m – a mass of the projectile, S – an area of projectiles middle section, ρ – an air density at the test conditions.

A coefficient calculated by this formula corresponds to the number of Mach $M_{av} = V_{av} / a$, where a – a local velocity of sound, $V_{av} = (V_1 + V_2)/2$.

Figure 6 shows the test results are grouped under the velocities. There are the results of aerodynamic tests [4–7] of the identical finned projectile layout obtained on the TSU model aerodynamic installation and data calculated using the aerodynamic drag law of 1958 year.

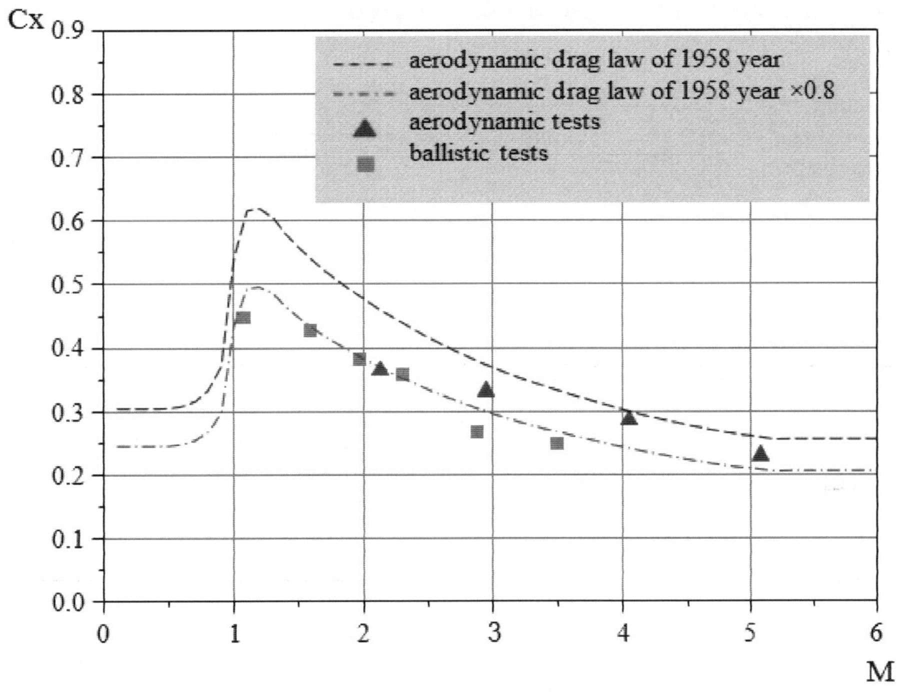

Figure 6. Comparison of test results finned projectile layout with the data of aeroballistic tests and the resistance law of 1958 year.

4. Conclusion

The nature of the obtained dependences generally repeats the resistance law of 1958 year with shape factor of 0.8.

Thus, the developed experimental ballistic track can be used to determine the drag coefficient projectile layouts in order to assess the extent of their target trajectory in the shot.

Acknowledgments

[1] This research carried out in 2017 was supported by «The Tomsk State University Academic D.I. Mendeleev Fund Program», grant № 8.2.05.2017.

References

[1] Konovalov A.A., Nikolaev Y.V. 1979 External ballistics. Central Research Institute of Information (Moscow)
[2] The application on patent RF No 2015127042. Muzzle velocity sensor / Burkin V.V., Diachkovskii A.S., Egorov A.L., Ishchenko A.N., Korolkov L.V., Maistrenko I.V., Rogaev K.S., Stepanov E.Y., Chupashev A.V. – Filing date: 06.07.2015
[3] Bimatov V.I., Merzlyakov V.D., Stepanov V.P., Faraponov V.V. 1981 Workshop on aeroballistics. Publishing house of Tomsk State University
[4] Zvegintsev V.I. 2014 Gasdynamic installations of short-term action. Parallel, Novosibirsk
[5] Maslov E. A., Faraponov V. V., Zolotorev N. N., Chupashev A. V., Matskevich V. V., Chizhov S. Yu. 2017 MATEC Web of Conferences **92**, 01056
[6] Maslov E.A., Zharova I.K., Kozlov E.A., Faraponov V.V., Savkina N.V., Zolotorev N. N., Chupashev A. V. 2017 MATEC Web of Conferences **115**, 02012
[7] Maslov E.A., Savkina N.V., Zharova I.K., Faraponov V.V., Kozlov E.A., V.V. Matskevich 2017 MATEC Web of Conferences **110**, 01055

Experimental studies of the physical process for forming the profile of a friction unit in a ballistic installation

A V Guskov[1], M A Kormushkin[2], K E Milevskii[3]

[1]Doctor of Technical Science, Professor, Novosibirsk State Technical University (NSTU), 20 Karl Marx street, Novosibirsk, 630073, Russia.
[2]Student, Novosibirsk State Technical University (NSTU), 20 Karl Marx street, Novosibirsk, 630073, Russia.
[3]Candidate of Technical Science, assistant professor, Novosibirsk State Technical University (NSTU), 20 Karl Marx street, Novosibirsk, 630073, Russia.

E-mail: a.guskov@corp.nstu.ru

Abstract. In the paper, physical processes were identified and investigated when the device being thrown in a ballistic installation moved. A set of methods for studying the processes that occur during the operation of the friction unit of the device being thrashed was developed. The process of plastic deformation during the formation of a friction unit of a device being thrashed, made of high-carbon steel, was studied. The process of forming a profile on the friction unit was studied, and the influence of the geometric parameters of the profile of the inner surface of the ballistic installation on the specific pressure of the formation of the friction unit of the device being thrown was determined. On the basis of practical experiments, conclusions were drawn about the effect of the geometric parameters of the friction unit profile on the punching force.

1. Introduction
The friction unit is an important element for the development and design of a projectile to be projected in a high-speed ballistic installation. The friction unit of the device being thrashed performs two priority tasks:

- Providing forward and rotational motion of the device being project at the expense of the profile of the inner surface of the ballistic installation;
- Sealing of powder gases.

The process of improving and modernization of the projected device was characterized by the necessity to retain the geometric parameters of the friction unit (diameter, width, and distance from the bottom) and performance properties: the resisting forces of the friction unit and the endurance of the ballistic installation.

When increasing the efficiency of the device, the retention of the properties requires multiple full-scale tests.

2. Goals and objectives of the experiment
Projectile is to identify and study the physical processes that occur when the device is being projected.

In the study, the main physical processes occurring in the ballistic installation during sample throwing were identified:

- Plastic deformation of the material of the friction unit while moving along the profile of the ballistic installation;
- Heat released in the interaction zone of the friction unit of the device being projected and the inner surface of the ballistic installation.

The isolated physical processes occurring during the operation of the friction unit of a missile are common and can be used to study the working capacity and the development of physical principles for designing friction units of missiles of all types, diameters and materials.

The procedure for studying the processes occurring at the stage of plastic deformation of the material of the friction unit of a missile consisted of a complex of experiments:

- Cavity formation on the friction node by the method of biaxial compression;
- Thrusting of the missiles with the friction nodes under investigation through the initial section of the internal profile of the ballistic installation;
- Deformation and destruction of the protrusions on the friction node.

3. Research procedure

The experimental method of biaxial compression based on the stepwise deformation of a rectangular projection of a friction unit of a new type by a punch with a working surface corresponding to the internal profile of a pipe of a typical ballistic installation was used as the primary method for investigating the plastic deformation of the material of the friction unit during incision (forming a depression between the protrusions).

For the formation of the cavity, ring segments with friction nodes made of typical preforms of missiles were used as samples. In addition, standard samples with variants of the friction unit of a new type with various geometries of a single protrusion from blanks of a missile, namely high-carbon steel, were manufactured.

The research consisted of three stages. At the first stage, a range of specific contact pressures *q* was established for typical friction assembly designs made of the following materials: iron ceramics, copper-nickel alloy and copper. The level of specific contact pressures *q* was:

- For the iron ceramics 1000-1100 MPa;
- For the Cu-Ni alloy from 1200 to 1280 MPa;
- For copper from 1150 to 1200 MPa.

It is necessary to pay attention to commensurability of the values of the specific contact pressures of the primary reaction of the friction node for the materials under study. The form of the dependence of the specific pressure on the degree of deformation in biaxial compression is shown in Figure 1 and indicates three sections in the formation of the projection.

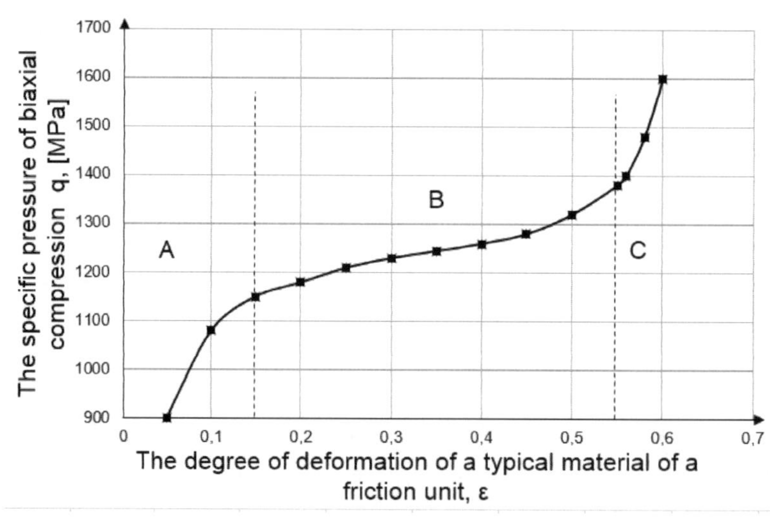

Figure 1. Dependence of the specific pressure of biaxial compression on the degree of deformation of a typical material of a friction unit $q = f(\varepsilon)$.

- Section «A» - the period of selection of gaps in the installation and the elasto-plastic deformation of the material of the friction unit in the range of ε from *0* to *0.15*;
- Section «B» - formation of the knot projection at ε from *0,15* to *0,55*;
- Section «C» - for $\varepsilon > 0.55$, the deformation of the material compression of the friction unit along the bottom and the cut field with a significant increase in the specific pressure.

A significant increase in the specific pressure in section «C» confirmed the need to limit the coefficient of deformation of the material of the friction unit when designing a new type of friction unit $\varepsilon \leq 0.55$. At the second stage of the research, a series of experiments was performed to establish the dependence of the specific pressure $q = f(\varepsilon)$ on the degree of deformation for the design of the friction unit of a new type. The degree of deformation ε of the protuberance material uniquely determines one of the main parameters of the friction node of a new type - the height of the protuberance, since $h = y_{max}/\varepsilon$. The maximum value of y_i corresponds to the height of the cut of the ballistic installation ($y_{max} = \Delta$). Therefore, to investigate the effect of the degree of deformation ε on an important indicator of the efficiency of a new type of friction unit, the specific contact pressure is necessary in a wide range of variation in the width and height of the single projection. Experiments on biaxial compression were carried out on samples with dimensions of a single protuberance with a width *c* from 1.4 to 2.0 mm and a height *h* from 1.8 to 3.6 mm. The value of the specific pressure q according to the method was determined as the ratio of the force on the punch to the actual contact area. The dependences $q = f(\varepsilon)$ (Figure 1) are similar for investigations of ring samples of friction units of a new type with a single projection and correspond to the dependences of the specific pressure on the degree of deformation for the friction nodes.

Only 5 series of samples of friction units of a new type with different ratios *h/c* of a single projection were tested. The accuracy of the contact pressure assessment is ± *8%* when the yield strength σ_{02} of high-carbon steel varies from 350 to 440 MPa.

Based on the results of statistical processing of experiments in the selected range of variation in the degree of deformation (0.30 < ε ≤ 0.55), the specific pressure was approximated from the degree of deformation for section «B»:

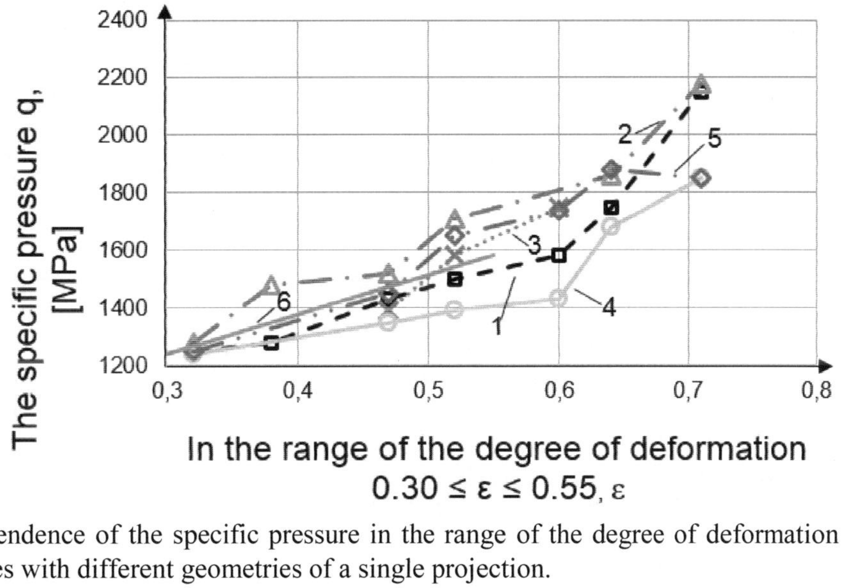

Figure 2. Dependence of the specific pressure in the range of the degree of deformation 0.30 ≤ ε ≤ 0.55 for samples with different geometries of a single projection.

$$q = 1500\varepsilon + 825 \tag{1}$$

On the basis of experimental dependences, an analysis of the magnitude of the operation of a biaxial compression characterizing the energy expenditure on the plastic deformation of the material of the friction unit of the missiles was made. The value of work of the biaxial compression of the single projection AC is about 96 J.

In the third stage, experiments were carried out on the biaxial compression of the single projection of the friction assembly on a specimen 122 mm in diameter. Based on the results of the experiments, the dependence of the force on the penetration depth of the cut was constructed: $P = f(y)$. During processing, the dependence of the specific contact pressure on the degree of deformation $q = f(\varepsilon)$ was obtained.

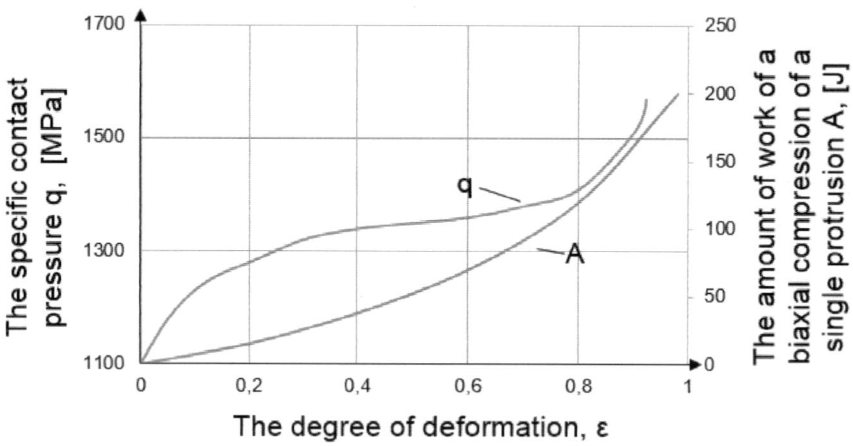

Figure 3. Dependences $q = f(\varepsilon)$ and $A = f(\varepsilon)$ obtained on a single projection of a new type of friction unit.

The shape of the curve $q = f(\varepsilon)$ indicates the presence of a section «B» in the biaxial compression process, in which the specific pressure is linearly dependent on the degree of deformation and corresponds to the dependence $q = 1500\varepsilon + 825$(1). In this section, the increase in the specific pressure is insignificant with respect to sections «A», «C».

When pressing the working part of the punch into the material of the friction unit, the deformation force increases. The increase in the specific pressure of the material resistance to the plastic flow at the time of completion of the formation of the depression with respect to the specific pressure at $\varepsilon = 0.5$ *is 1.23* times.

The second step in the experiment is to determine the loads when the missiles are pushed through the profile of the ballistic installation.

Investigation of the process of plastic deformation of the material of the friction unit of a new type and determination of the punching force was carried out on a section of a profile pipe 122 mm in diameter and with dimensions corresponding to a typical ballistic installation.

The conical part of the tube of the ballistic installation, which performs the function of the stamping tool at the initial moment of motion of the projected sample, consists of a smooth conical part and a conical slope of the rifling. When introducing and sending the projectile, the friction unit is fixed. The obtained compound ensures the preservation of the propellant gases at the initial moment of motion.

The process of forming protrusions on the friction node includes two stages:

- Deformation of the material of the friction unit;
- Formation of protrusions on the friction unit.

To determine the axial force and the distribution of specific pressures, a technique based on experiments at the selected stages was developed. To optimize manufacturing, a batch of samples consisting of three groups was prepared: two groups with friction nodes of a new type with different ratios *h/c*, a third control group with a copper friction unit.

To determine the dependence of the punching force on the geometric parameters of the construction of a friction unit of a new type, studies were carried out on a batch of samples with a diameter of 45 mm. All samples were made of high-carbon steel.

The dependence of the punching force $P_1 = f(c)$ for different hub height h is shown in Figure 4. The value of P_1 is inversely proportional to the height of the protuberance *h*, which essentially depends on the change in the width of the projection *c*. Thus, for $h = 1\ mm$, a change in the width of *c* by a factor of 5 leads to an increase in the force by 6.5 times, and at $h = 2\ mm$ only in 2.7.

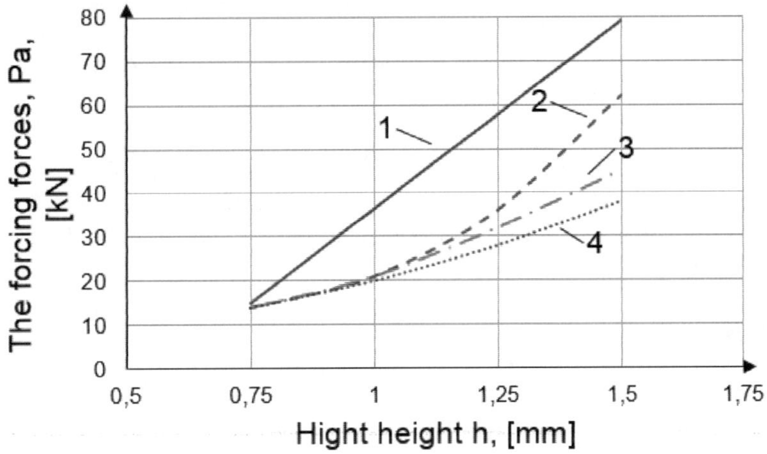

Figure 4. – Dependence $P_1 = f(c,h)$

With the range of changes in the *c/h* ratio from 1.5 to 0.375 (4 times), the axial punching force is significantly reduced from 79.3 to 13.3 kN (about 6 times). This result confirmed that the friction unit of a new type is a design that allows to regulate the interaction force of a node with a ballistic installation by changing the geometric parameters of the projections and the distance between them. In this case, the width of the projection defines the working area of the projection. This must be taken into account when specifying admissible deviations to the size of *c* when manufacturing the friction unit.

To confirm the results obtained on samples of 45 mm, the extrusion force from the parameters of the friction unit of a new type (*h* and *c*), the width *B* (the number of protrusions) and the diameter D of the friction unit, a pilot batch of 122 mm diameter samples was produced. The protrusion of samples through a section of the profile pipe was carried out in two stages.

At the first stage, a free movement and rotation of the sample occurred in the connecting cone, the cone of penetration. The material yield strength σ_{02} of the missiles varied from 355 to 460 MPa.

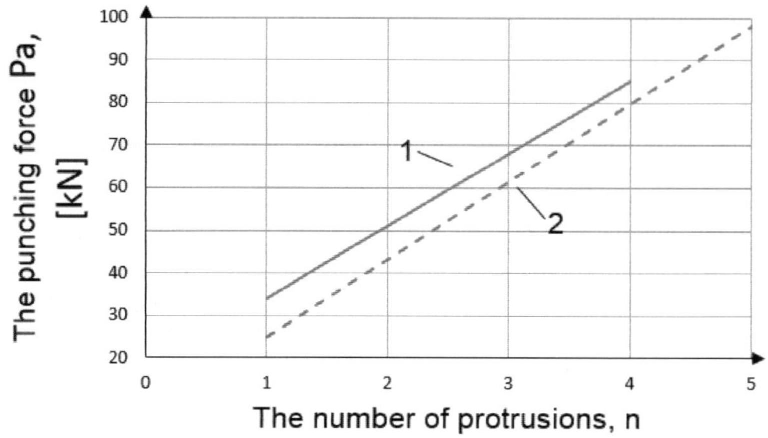

Figure 5. Dependence of the punching force on the number of protrusions

The values of the coefficients at *h, c, b, D* determined the degree of influence of the corresponding parameter of the friction unit of a new type on the variation of the axial force P_1 and confirmed the determining role of the height and width of the protuberance in the interaction of the new type of friction unit with the profile tube of the ballistic installation during the formation of the protrusions.

4. Results and discussion

Research based on the method of biaxial compression made it possible to conclude that the geometric parameters of the friction unit of a new type and the specific pressure necessary to form the protrusion are interconnected by one of the main parameters of the rifled portion of the ballistic installation, the depth of the cut Δ.

The following dependencies are established through the profile when estimating the loading of the missiles to be extruded through the profile:

- The force of extrusion of samples with a new type of friction unit through a pipe with a smooth internal surface is less than 1.7 times the copper nodes;
- The force in the second stage when forming the protrusion exceeds the force at the first approximately 1.55 times for the friction unit of a new type and 2.2 times for the brass.

When the angles of inclination of the front cone of a smooth pipe and the conical slope of the profile pipe are equal, the rolling force is proportional to the pressure in the contact "profile pipe-friction unit", i.e., the specific pressure.

5. Conclusion

According to the results of the experiments (Figure 5), it is determined that the dependencies of the force of extrusion depend on the number of protrusions. With an increase in the number of protrusions, the punching force increases significantly. It has been established experimentally that the width of the projection of the friction unit exerts the main influence on the punching force.

References

[1] Krafft I M 1995 Surface Friction in Ballistic Penetration *J. Appl. Phys.* **26** pp 81-94
[2] Earles S W and Kadhim M I 1966 *Friction and wear of unlubricated Steel surfaces and speeds up to 655 ft/s.* Instr. Mech. Engrs. vol. 180, Pt. 1, N. 22. 22
[3] Balakin V A 1980 *Friction and wear at high sliding speeds* Moscow p 136.
[4] Gusachenko L K, Guskov A V, Ivania S P and Milevskii K E 2011 *Modern problems of mathematics and mechanics* (Tomsk: Tomsk State University press) pp. 343–348

[5] Guskov A V, Zeziulinskii J S and Milevskii K E 2016 *Numerical simulation of the temperature distribution in the material of the steel node* (Tomsk: Tomsk State University press) p 118
[6] Gerasimov A V 2007 *Theoretical and experimental studies of high-speed interaction of solids* (Tomsk: Tomsk State University press) p 514–525
[7] Guskov A V, Zeziulinskii J S, Volf N A, Vandakurov K A and Milevskii K E 2017 *A new type master belt of an artillery projectile for rifled weapons* 18th Russian Science and Technical conf. (Novosibirsk: Novosibirsk State Technical University press) vol 2 pp 38-41
[8] Guskov A V, Zeziulinskii J S and Milevskii K E 2016 *Heat and mass transfer in the system of thermal modes of energy – technical and technological equipment* vol 72 MATEC Web of Conferences
[9] Guskov A V, Milevskii K E and Khrebtova M S 2016 The new type of leading band *Bulletin of the Academy of Military Sciences* **41** pp 22-25
[10] Guskov A V, Zeziulinskii J S Milevskii K E and Chaika N F 2015 Heat transfer in high-speed model leading belt for channel gun mount *GEO-Siberia-2015* **3** pp 107-111

Detection of structural changes and mechanical properties of light alloys after severe plastic deformation

V A Krasnoveikin[1,2], A A Kozulin[1] and V A Skripnyak[1]

[1] National Research Tomsk State University, Lenina ave. 36, Tomsk 634050, Russia
[2] Institute of Strength Physics and Materials Science of Siberian Branch Russian Academy of Sciences, pr. Akademicheskii 2/4, Tomsk 634055, Russia

E-mail: volodia74ms@yandex.ru

Abstract. Severe plastic deformation by equal channel angular pressing has been performed to produce light aluminum and magnesium alloy billets with ultrafine-grained structure. The physical and mechanical properties of the processed alloys are examined by studying their microstructure, measuring microhardness, yield strength, and uniaxial tensile strength. A non-destructive testing technique using three-dimensional X-ray tomography is proposed for detecting internal structural defects and monitoring damage formation in the structure of alloys subjected to severe plastic deformation. The investigation results prove the efficiency of the chosen method and selected mode of producing ultrafine-grained light alloys.

1. Introduction

Over the past decades, intensive efforts have been made to develop new technologies with the use of severe plastic deformation (SPD) methods for the production of ultrafine-grained (UFG) and nanostructured (NS) Al- and Mg-based light alloys. This has provided a qualitative increase in the efficiency of automotive and aerospace engineering, ship building, medical implants and equipment, power generation systems, and sports gear. The main aspects of the SPD influence on the properties of structural metals and alloys have been noted by many researchers [1-4]. They consist in the refinement of the material internal structure to UFG and NS states, due to which the physical and mechanical properties of the material change and the material hardens. The hardening effect after SPD shows up as an increase in microhardness, yield strength and ultimate tensile strength, a change in the degree of specimen elongation to failure, and an increase in durability and fatigue strength [4, 5]. There are different methods of SPD in metals and alloys that can be performed in local zones, on surfaces, and in the bulk of the material [4, 6-9]. The most common method is equal channel angular pressing (ECAP) that allows changing the structure in bulk specimens. A correctly selected scheme and an optimal pressing mode should provide uniform microstructure refinement in alloys without the formation of micro- and macrodefects and without cracking of the specimen material.

The major goal of this study is to define the effect of SPD on the physical and mechanical properties of the Russian light structural alloys such as aluminum alloy of grade 1560 and magnesium alloy of grade Ma2-1, which are the closest analogues of the known alloys 5083 and AZ31, respectively. These alloys are widely used in aviation, automotive and space industries. By changing their physical and mechanical properties, it becomes possible to optimize structural elements and to increase the service efficiency under various operating conditions owing to the reduction in the weight of parts and structures.

2. Materials and methods

Prismatic billets (measured 8x8x40 mm^3) for processing and subsequent investigation were prepared from bars of hot rolled coarse-grained alloys 1560 (chemical composition: 91% Al – 6.12% Mg – 0.59 Mn - ~ 2.29% other impurities) and Ma2-1 (chemical composition: 93.65% Mg – 4.36% Al - 1.33%

Zn – Mn - ~ 0.66% other impurities). The main billet axes coincided with the bar rolling direction. The billets were subjected to SPD by ECAP to produce UFG structure using a collapsible steel die (figure 1a) with the channels intersecting at an angle of 90 degrees without fillets. The given die geometry provides the maximum plastic strain accumulation in the material in one pass, equal to 115%. The pressing was carried out using route B_c (figure 1b) in which the billet is rotated by 90 degrees about the longitudinal direction (LD) in every next pass. The chosen ECAP scheme was assumed to be the most suitable for producing a homogeneous UFG structure within the specimen and to provide isotropic mechanical properties [4]. In order to eliminate defects in the ECAPed billets of the both alloys, optimal pressing modes were selected by changing the governing experimental parameters. Pressing was carried out on a universal servo-hydraulic testing machine INSTRON 40/50-20 equipped with an oven which permits a fine adjustment of pressing speed and force up to 50 kN.

(a)

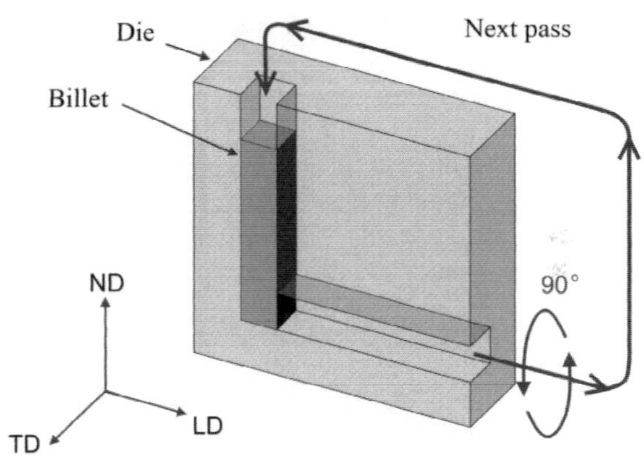

(b)

Figure 1. General view of the collapsible die (a) for ECAP, and (b) schematic drawing of route B_c, here values LD, ND and TD are longitudinal, normal and transvers directions, respectively.

This paper reports the X-ray tomography results on the detection of internal and surface defects in the alloys after processing. The quality of the processed billets was assessed by non-destructive testing using the microfocus X-ray system YXLON Y. Cheetah [10,11].

A series of tests were performed on the processed prismatic billets to examine the physical and mechanical properties of the studied materials, including microstructural analysis, microhardness measurements, and uniaxial tensile testing of flat specimens.

The texture and microstructure of the material before and after processing were observed by EBSD analysis using a scanning electron microscope Tescan Vega II LMU equipped with an EBSD device. The specimen surface was prepared for the investigation by mechanical polishing and ion etching using an ion slicer EM-09100. The obtained data were analyzed with the licensed HKL-Channel 5 software.

The mechanical texture of alloys after SPD was examined on an optical inverted microscope Olympus GX. The gage part of the flat specimens produced for uniaxial tensile tests was chemically etched for optical microscopy.

The Vickers microhardness (HV) measurements were conducted according to ISO 6507-1:2005 on the lateral surface of the prismatic billets in different regions along the pressing axis using a microhardness tester HMV G21ST Shimadzu. The measurements were taken under a load of 0.5 N and a dwell time of 10 s. Over 100 measurements were made for each specimen. The specimen surfaces were prepared by standard techniques using mechanical grinding and polishing to a defect-free mirror finish.

Flat dumbbell specimens for tensile testing with the gage part measured 10 mm in length, 1.6 mm in thickness, 3 mm in width, and with a 2.5 mm shoulder radius were cut out from the ECAPed prismatic billets along the pressing axis by electric discharge machining. Uniaxial tensile tests were conducted on a servo-hydraulic testing machine INSTRON 40/50-20 with the strain rate 0.001 1/s and test temperature 298 K. The tensile loads applied to the specimens were recorded with an accuracy of 0.05% using a load cell Dynacell.

3. Results and Discussion

While selecting optimal pressing modes for the studied materials, some crucial governing parameters were experimentally revealed which must be taken into account. The first parameter is the temperature, which must be 473 K for aluminum alloy and 523 K for magnesium alloy. The second parameter is the pressing speed, which was equal to 15 mm/min. The third parameter is friction. Friction between the billets and the die channel walls was minimized using a high-temperature lubricant on the basis of mineral oil and molybdenum disulphide [12]. With the above parameters, the processed material preserves its continuity without cracking.

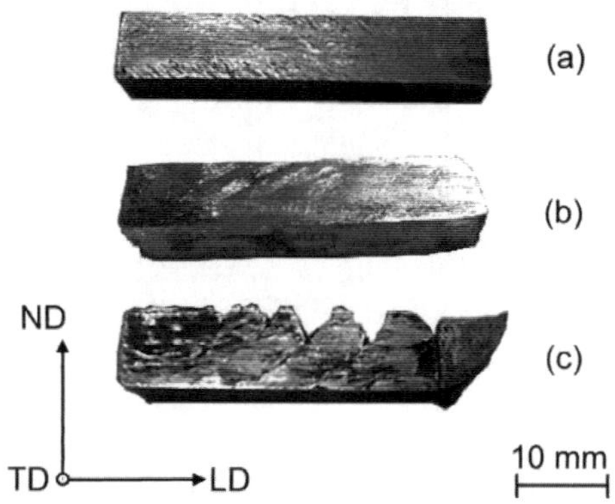

Figure 2. Prismatic billets before and after pressing: (a) – as-received; (b) – after processing by the optimal ECAP mode; (c) – fragmented specimen after processing by the non-optimal ECAP mode.

The fragmented billet depicted in figure 1c is an example of using a non-optimal pressing mode. Cracks that cause the billet fragmentation are generated at an angle of 45 degrees to the longitudinal direction along the maximum shear strain direction in the channel intersection zone. The billet processed in the optimal mode (figure 2b) shows no visible signs of cracking.

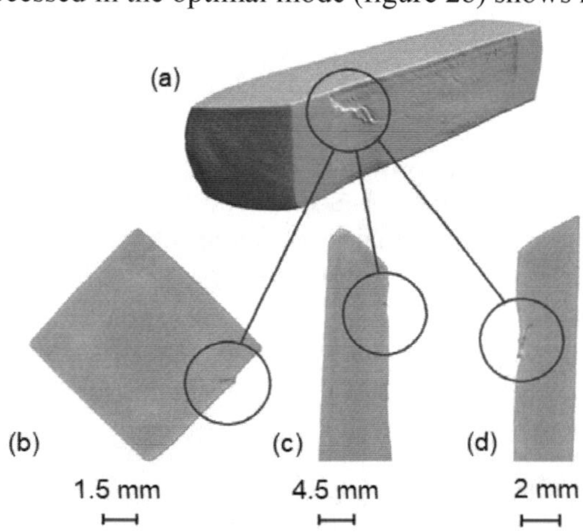

Figure 3. Voxel model of prismatic billets subjected to SPD: isometric projection (a); front section (b); left (c) and right (d) longitudinal section.

Modern X-ray tomography methods were applied to evaluate the internal structure, porosity, the presence of microcracks and voids in the bulk of the ECAPed billets without visible defects. X-ray tomography provides a 3D image of an object in the form of a voxel model that allows viewing an image of any surface or internal section with micron accuracy [10]. Figure 3 shows a voxel model of the prismatic billet with a crack after ECAP processing. The model is shown in an isometric projection and three sections. The surface defect viewed as a small step is a crack propagating from the surface into the bulk. The billets which had such defects were considered unsuitable for further investigation and were rejected. The X-ray tomography results confirmed that if SPD is performed in the optimal mode, the structure of the material after processing is free from macro- and microdefects.

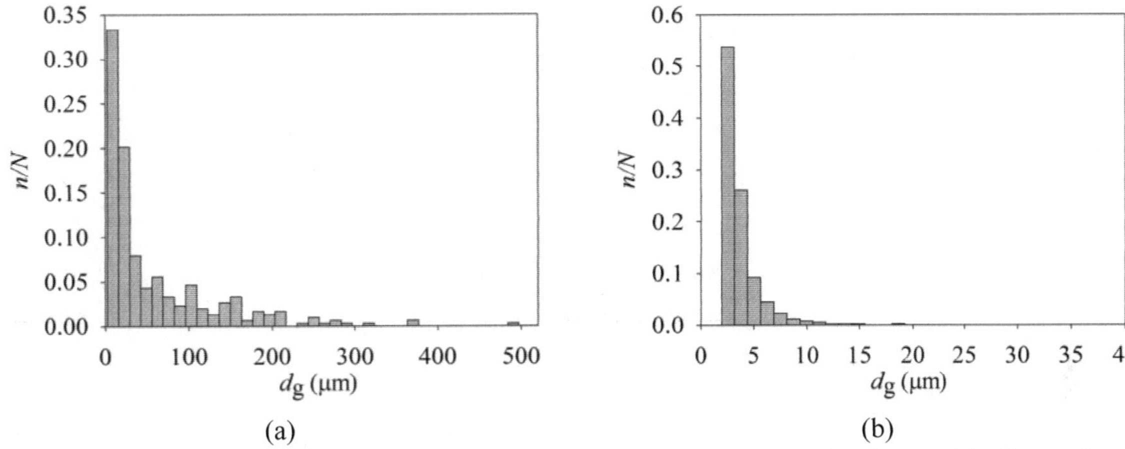

Figure 4. Grain size distribution in aluminum alloy 1560: (a) –as-received material; (b) – after 4 ECAP passes.

Figures 4 and 5 demonstrate grain size distribution histograms, where d_g is the grain size in μm, N and n is the total number of measurements and the number of grains of a certain size, respectively. Microstructural studies has shown that aluminum alloy 1560 in the as-received state exhibits a grain size distribution within a wide range from 2 to 400 μm, with the average grain size 50 μm. The grain size distribution in the as-received magnesium alloy Ma2-1 ranges from 2 to 60 μm, with the average grain size 18 μm. Since the billets were cut out from hot rolled bars, the grain structure has a well-defined direction. Coarse grains are elongated along the rolling direction and clusters of fine grains are adjacent to them, together forming bimodal structures.

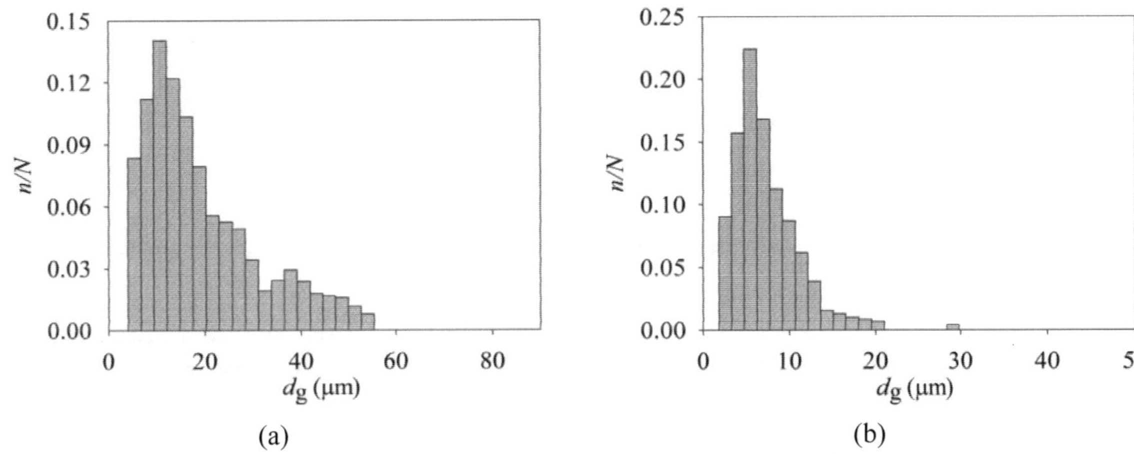

Figure 5. Grain size distribution histograms in magnesium alloy Ma2-1: (a) –as-received material; (b) – after 4 ECAP passes.

After four ECAP passes in the selected optimal mode, a more homogeneous UFG structure with an average grain size of 3 and 7 μm is formed in the central part of aluminum and magnesium alloy billets, respectively.

Figure 6 illustrates the averaged experimental data obtained in Vickers microhardness (HV) measurements on the studied materials before and after ECAP. Here, the value $\Delta d/d_0$ is the ratio between the distance from the measurement area to the billet edge Δd (mm) and its total length d_0 (mm). Statistical analysis for all results was carried out by sampling and averaging the obtained data, with the determination of the confidence intervals.

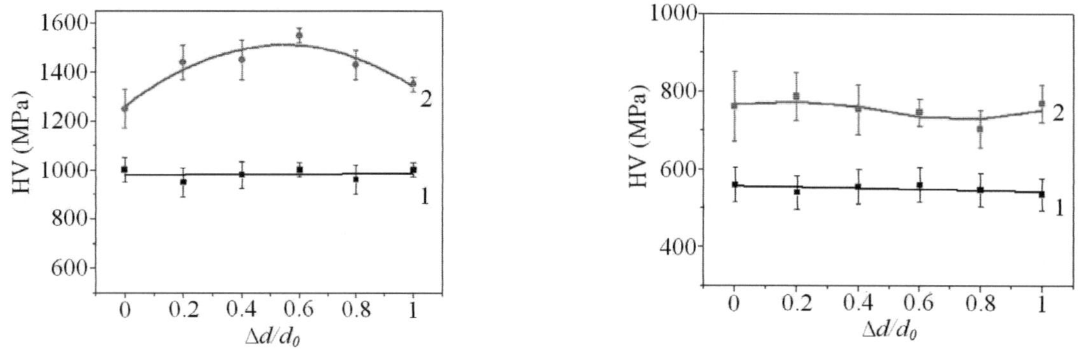

(a) (b)

Figure 6. Comparison of microhardness measurement results for aluminum alloy 1560 (a) and magnesium alloy Ma2-1 (b) along LD - direction: 1 – as-received material; 2 – ECAPed material.

Experimental results indicate that ECAP leads to an increase in microhardness in the entire specimen volume, but there is inhomogeneity in the axial direction. After four ECAP passes, the microhardness increases by on average 40–60 % as compared to its initial value for the both materials. The maximum microhardness reaches 1550 and 786 MPa, with the initial value not exceeding 1000 and 560 MPa for the aluminum and magnesium alloys, respectively. The microhardness in the head ends of aluminum specimens is somewhat lower (up to 10 %) than the average value. The microhardness deviation from the average values in the central specimen part both along the longitudinal axis and in the transverse direction is within 10 %.

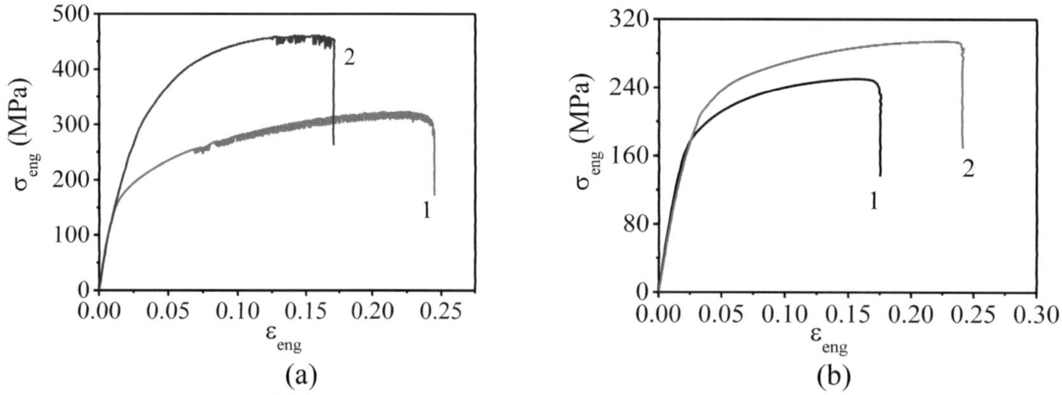

(a) (b)

Figure 7. Uniaxial tensile test results for specimens of aluminum alloy 1560 (a) and magnesium alloy Ma2-1 (b): 1 – as-received material; 2 – ECAPed material.

Uniaxial tensile test results are represented as stress-strain curves in figure 7. It is seen that after four ECAP passes the yield strength of aluminum alloy 1560 increases from 150 to 270 MPa and its ultimate tensile strength increases from 320 to 460 MPa, while those of magnesium alloy increase from 150 to 200 MPa and from 250 to 290 MPa, respectively. The maximum elongation to failure under uniaxial tension increases from 0.17 to 0.24 for magnesium alloy and decreases from 0.24 to 0.17 for aluminum alloy. These data indicate that the strength characteristics in the bulk of the both studied alloys increase during SPD by ECAP.

The serrated flow curve for uniaxially tensile aluminum alloy is indicative of serrated flow occurring due to the Portevin-Le Chatelier effect. The serrated profile is not observed in the entire plastic flow region. It begins with 6% plastic deformation in the as-received alloy, and then the beginning of serration is shifted to higher strains up to 10% in the processed alloy with UFG structure.

The stress-strain curves for magnesium alloy in the as-delivered state and after ECAP are smooth.

4. Summary

The ECAP temperature and speed were determined experimentally for aluminum alloy 1560 and magnesium alloy Ma2-1. An optimal processing mode was selected to produce specimens without defects in the entire volume. A non-destructive testing technique on the basis of X-ray tomography was proposed which enables the detection of defects such as cracks and voids both on the surface and in the bulk of prismatic billets.

The investigation results showed that for SPD of light alloys 1560 and Ma2-1 by multipass ECAP four passes are enough to produce a homogeneous structure in the specimen bulk with an average grain size of 3 and 7 μm, respectively. The alloys processed in this mode assume an ultrafine-grained structure.

It was found experimentally that SPD of the studied alloys increases the yield strength and short-term strength. The maximum elongation to failure under uniaxial tension increases for magnesium alloy but decreases for aluminum alloy. The microhardness of the both alloys after four ECAP passes increases by 40–60%.

Acknowledgments

The work was supported by the RF President Grant MK-5914.2016.1. The authors are grateful to Prof. D.V. Lychagin for the permission to use the equipment of the Shared Use Analytical Center of Geochemistry of Natural Systems of Tomsk State University, and for discussion of results.

References

[1] Meyers M A, Mishra A and Benson D J 2006 *Prog. Mater. Sci.* **51** 427

[2] Sabirov I, Murashkin M Yu, Valiev R Z et al. 2013 *Mater. Sci. Eng. A.* **560** 1

[3] Markushev M V, 2011 *Letters on Mater.* **1**, 36

[4] Valiev R Z, Langdon T G 2006 *Prog. Mater. Sci.* **51 (7)** 881

[5] Krasnoveikin V A, Kozulin A A, Skripnyak V A and Moskvichev E N 2016 *VI International Scientific Conference «Current issues of continuum mechanics and celestial mechanics – 2016»* 135

[6] Gupta A K, Maddukuri T S and Singh S K 2016 *Prog. Mater. Sci.* **84** 403

[7] Lu K, Lu J 2004 *Mat. Sci. Eng. A.* **375-377** 38

[8] Bodrov A S, Olimpieva N V, Zelepugin A S and Zelepugin S A 2015 *Vestnik Tomskogo gosudarstvennogo universiteta. Matematika i mekhanika* **5** 37

[9] Moskvichev E N, Skripnyak V A, Skripnyak V V, Kozulin A A and Lychagin D V 2016 *Letters on Mater.* **6**(2) 141

[10] Narikovich A S, Leitsin V N, Kulkov S N 2016 *Transport and Services* **4** 79

[11] Kozulin A A, Narikovich A S, Kulkov S N, Leitsin V N and Kulkov S S 2016 *AIP Conf. Proc.* **1760** 020036

[12] Kozulyn A A, Skripnyak V A, Krasnoveikin V A, Skripnyak V V and Karavatskii A K 2015 *Russ. Phys. J.* **57**(9) 1261

IOP Conf. Series: Journal of Physics: Conf. Series **919** (2017) 012013 doi:10.1088/1742-6596/919/1/012013

Numerical modeling and prediction of mechanical properties of ceramic composite

V A Mikushina[1], **I Yu Smolin**[1,2] **and Yu N Sidorenko**[1]

[1] Tomsk State University, 36 Lenin avenue, Tomsk, 634050 Russia
[2] Institute of Strength Physics and Materials Science SB RAS, 2/4 Akademicheskii avenue, Tomsk, 634055, Russia

E-mail: mikushina_93@mail.ru

Abstract. The paper is devoted to the numerical simulation of the composite consisting of ZrO_2-based ceramics and cortical bone performed with making use of a multi-scale approach. The specificity of the approach considered is the using of a geometrical model of the composites representative volume that must be generated with taking into account the composite reinforcement structure. It is shown that such approach can be used to predict the value of composite macroscopic ultimate strength. The evolution of mesoscopic stress distributions in the composite components during its deformation was investigated with taking into account damage accumulation up to the fulfillment of the macroscopic strength criterion. It is shown that damage accumulation has an impact on the stress distribution laws at the mesoscopic level, which is manifested in the appearance of a threshold for the stress distribution, as well as in a significant decrease in distribution amplitude.

1. Introduction

Currently, modern medicine uses new materials for the production of high-quality medical equipment and development of new technologies. For example, products from ceramic materials can be used to produce a "substitute" of destroyed or damaged human tissues in surgical medicine. Ceramic materials have unique properties. These are high hardness, wear resistance, low density, high compressive strength, etc. For bone prostheses, the great interest is expressed in ceramic materials based on zirconia. These materials are characterized by high strength and excellent biocompatibility with living tissues of the human body [1–6]. A successful application of such materials in medicine, as in other applications, is largely determined by the completeness of information on their mechanical and physical properties. Computer simulation is widely used currently to obtain such information. The correctness of these simulations directly depends on the quality of computer models of the materials. Different numerical techniques and approaches can be used to evaluate the mechanical properties of ceramic materials. The authors of studies [7–11] used the movable cellular automaton method in modeling brittle porous ceramics. The advantage of this method is the successful modeling of multiple cracking and fracture of materials that was shown in the above mentioned papers. Here we are going to use the conventional finite element method like in [12–14]. Regarding the influence of the structure on the mechanical properties of composite materials, we follow one of the most promising approaches as in papers [7–18] and many others. It is a multi-scale approach. This approach allows taking into account the type of structure and its influence on the properties of composites.

The aim of the paper is to study the features of changes in the stress, strain and damage distributions at the mesoscale level in ceramic composite materials under mechanical loading. To achieve this goal numerical simulation of the mechanical behavior of the composite was performed with taking into account damage accumulation.

2. Numerical Model and Method

The composite material consists of matrix and filler. The matrix is porous ZrO_2-based ceramics. The filler material is bone tissue. So, it is a kind of biocomposite when the pores of ceramics are filled with bone tissue. The porous structure of the ceramic implant is described explicitly. We suggest that the pores have a circular shape. Their radius and position within the mesovolume geometrical model are determined in a random manner. The size of the square comprising the representative geometric model is 15 times larger than the greatest pore size, which was shown to be quite enough [19].

The study was conducted within the multi-scale approach to the description of mechanical properties of materials discussed in [12]. A composite material is considered at several scale levels: macro-, meso- and microlevels. On the macro level, a material is treated as an quasihomogeneous one. The properties of the composite material at this level may vary with space coordinates but quite smoothly. So, the properties at this level are characterized by some effective values representing the result of compilation of information about the behavior of the material obtained at the mesolevel. Mesovolume is a fragment of a piecewise-homogeneous medium for which the distribution of the components and loading conditions are clearly defined [20, 21]. For the mesovolume there is a concept of locally-effective properties, which are defined as the average (effective) ones for the selected limited fragment of a heterogeneous medium. On the micro level we consider the volumes which dimensions are much larger than the interatomic distances but at the same time are smaller than the characteristic dimensions of the composite structural components. It is believed that for such a volume the postulates of continuum mechanics are acceptable, i.e. the hypothesis of continuity and homogeneity hypothesis are accomplished for it, and mathematical apparatus of integral and differential calculus can be used for its description. The properties of the material at this level are described by constitutive relations of anisotropic media (e.g., crystalline or bone tissue components).

Mechanical behavior of the composite on all scale levels is governed by equations of solid mechanics which include equilibrium equations, Cauchy (geometric) equations and constitutive relations [22]. When the local failure criterion is fulfilled the parameters of mesovolume damage for each component of the composite are calculated according to the formula [23]:

$$\Pi_k = \frac{V_k^*}{V_k}, \tag{1}$$

where V_k is the total volume of the k-th component in the mesovolume, V_k^* is the damaged fraction of the volume of the k-th component in the same mesovolume. The modulus of elasticity is calculated taking into account the damage for each k-th component of the composite by the equation:

$$E_k = E_k^0(1-\Pi_k), \tag{2}$$

where E_k^0 is the value of the modulus of elasticity of the undamaged material. It is believed that the mesovolume retains its carrying capacity until the following condition is fulfilled for each of its structural components:

$$\Pi_k > \Pi_k^{\max}, \tag{3}$$

where $\Pi_k^{\max}$ is the maximum permissible value of damage in the k-th component of the composite. This value was chosen taking into account both physical considerations and the stability requirements of the computational method and according to the studies in [23] was considered equal to 0.75. To estimate the strength of material at the macro-scale we used a percolation criterion discussed in [23].

The calculation of stress and strain for the composite at the meso level was performed using the finite element method [24] in the two-dimensional formulation. Various examples of the investigated finite element models of mesovolumes are represented in figure 1. The case of quasi-static uniaxial loading along the vertical axis and elastic-brittle constitutive model of material were adopted.

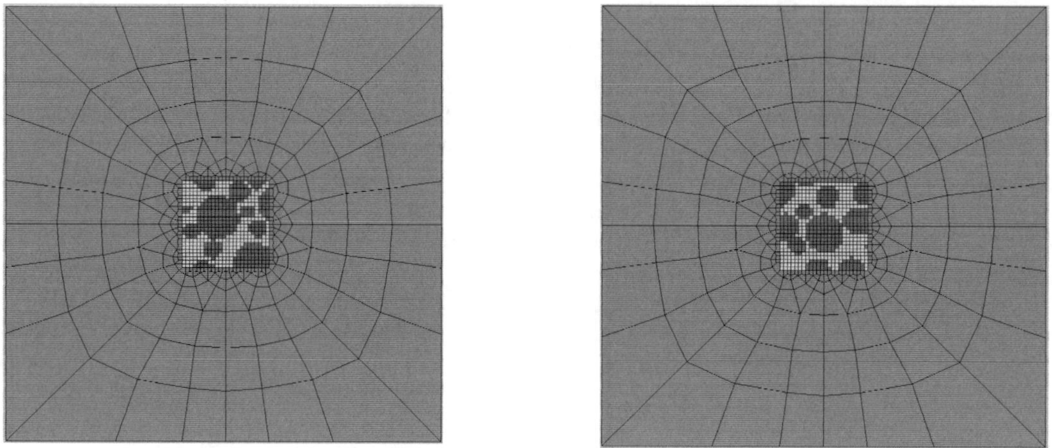

(a) (b)

Figure 1. Typical finite-element representations of characteristic fragments of the model structure.

3. The Results of Computer Modeling

As a result of the calculations, the parameters of the distribution laws of effective values of strain and stress at the mesolevel in the course of the material loading were determined. The evolution of the distributions of the mesoscopic stress in the components of the composite during its uniaxial compression with taking into account damage accumulation up to the fulfillment of the macro strength criterion was investigated and presented in figure 2.

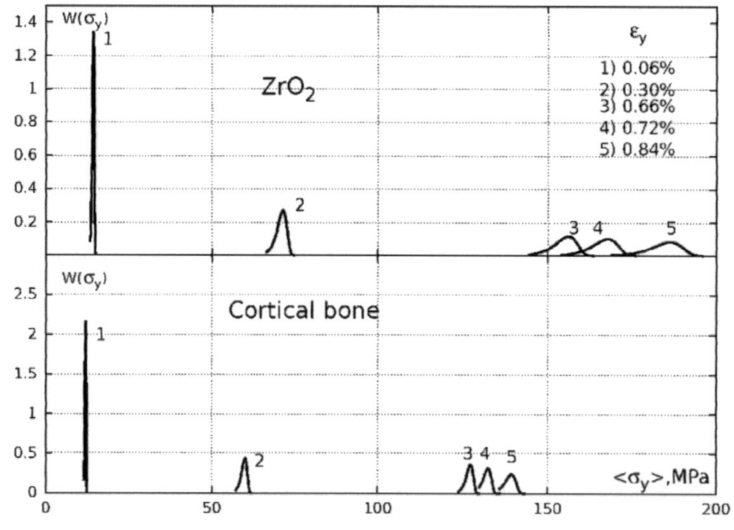

Figure 2. The Weibull probability density functions of the stress components along the axis of compression σ_y for ceramics and cortical bone at different values of effective strain.

The following values of strain were selected as characteristic ones for the analysis of evolution of stress distributions: 0.06 % and 0.3 % for the elastic stage of deformation; 0.66 % for the onset of damage accumulation; 0.72 % for the formation of a connective cluster of damages, 0.84 % for the stage of extensive damage in the mesovolume. The Weibull distributions are shifted to higher stresses with an increase in strain for ceramics and bone tissue. The scatter of the distributions at the initial stage of loading is extremely small as can be seen in figure 2 that points to the uniformity of stress fields in this case. The scatter increases with the growth of load implying the increase of stress field irregularity that is explained by damage accumulation and different local values of the modulus of elasticity according to equation (2).

The stress distribution functions are similar for each component of the composite before damage accumulation starts as one can see in figure 3a. The similarity of distribution laws can be explained by the fact that in the absence of damage the modulus of elasticity of each component keeps its value and all the stresses increase proportionally with increasing strain.

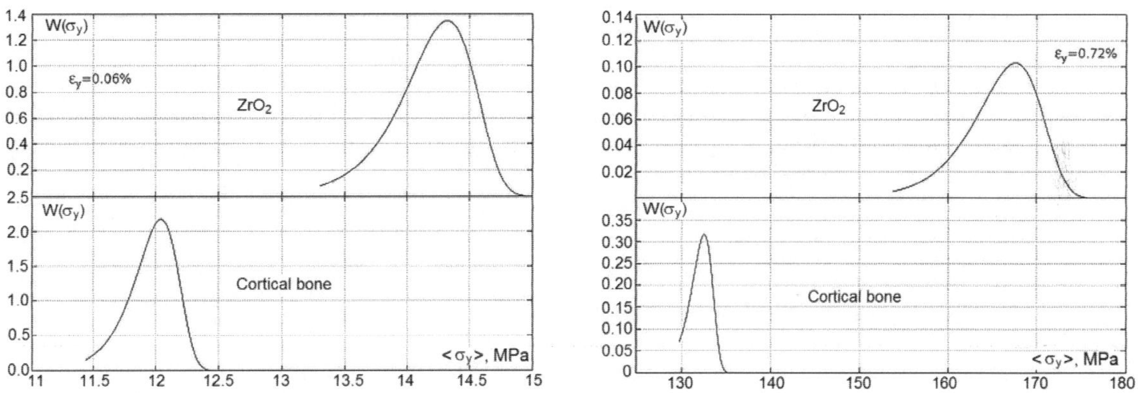

Figure 3. The Weibull probability density functions of the stress components σ_y for ceramics and cortical bone at the following values of effective strain 0.06 % (a) and 0.72 % (b).

In the presence of damage, the stress distributions for ceramics remain similar to the stress distributions in the absence of damage though the scatter increases (figure 3b). For the bone tissue, the scatter of stress distributions is smaller as compared to the ceramics but the parameters of stress distribution laws change as compared to the intact condition, in particular, a lower threshold of distributions is observed as in figure 3b. The reason of this is more active damage accumulation in the bone tissue in comparison with the ceramic matrix.

4. Summary

The study of the evolution of the mesoscopic stress distribution laws in the composite components with taking into account damage accumulation showed that stress distributions for different values of strains are similar to each other in the absence of damage, but the scatter of the distributions increases and its amplitude reduces with the increase in strain. It is shown that damage accumulation has an impact on the stress distribution laws at mesoscopic level, which is manifested in the appearance of a threshold for the stress distribution, as well as in a significant decrease in distribution amplitude.

The value of the effective Young's modulus of the studied composite materials was obtained to be $E = 18.7$ GPa which is quite close to the Young's modulus of bone tissue $E = 16$ GPa. The limit values of effective stress and strain corresponding to percolation criterion of strength are found to be $\sigma_{eff} = 124$ MPa, $\varepsilon_{eff} = 0.68$ %.

Acknowledgments
The research was supported by "The Tomsk State University competitiveness improvement programme", grant 8.2.12.2017 and the Program of basic scientific research of the Russian Academy of Sciences in 2013–2020, project 23.2.3.

References
[1] Kanukov V N, Strekalovskaya A D and Kilkinov V I 2004 *Materials for Modern Medicine* (Orenburg: Orenburg State University Press) p 113 (in Russian)
[2] Dubok V A, Protsenko V V, Shinkaruk A V and Atamanenko O N 2008 A new generation of bioactive ceramics: peculiarities of their properties and clinical results *Orthopaedics, Traumatology and Prosthetics* **3** 91–5 (in Russian)
[3] Mikhailina N A, Podzorova L I, Rumyantseva M N, Shvorneva L I., Ovchinnikova O A,. Anisimova S V, Lebedenko I Yu, Lebedenko A I and Khvan V I 2010 *Inorg. Mater. Appl. Res.* **1** 335–8
[4] Saini M, Singh Y, Arora P, Arora V, Jain K 2015 *World J. Clin. Cases* **3** 52–7
[5] Maccauro G, Rossi Iommetti P, Manicone P and Raffaelli L 2012 *Zirconia and Alumina Bioceramic Biocompatibility* (New York: Nova Science Publihsers)
[6] Afzal A 2014 *Mater. Express* **4** 1–12
[7] Smolin A Yu, Roman N V, Konovalenko Ig S, Eremina G M, Buyakova S P and Psakhie S G 2014 *Eng. Fract. Mech.* **130** 53–64
[8] Smolin A Yu, Shilko E V, Astafurov S V, Konovalenko I S, Buyakova S P and Psakhie S G 2015 *Def. Technol.* **11** 18–34
[9] Dmitriev A I, Smolin A Yu, Popov V L and Psakhie S G 2009 *Phys. Mesomech.* **12** 11–9
[10] Konovalenko Ig S, Smolin A Yu and Psakhie S G 2015 *AIP Conf. Proc.* **1683** 020089
[11] Vertyagina Y and Marrow T J 2017 *Carbon* **121** 574–90
[12] Mikushina V A and Sidorenko Yu N 2015 *AIP Conf. Proc.* **1683** 020150
[13] Lasko G, Weber U, Apel M, Berger R and Schmauder S. (2015) *Adv. Eng. Mater.* **17** 1536–46
[14] Soppa E, Nellesen J, Romanova V, Fischer G, Crostack H-A, Beckmann F 2010 *Mater. Sci. Eng.* A **527** 802–11
[15] Makarov P V, Smolin I Yu, Prokopinsky I P and Stefanov Yu P 1999 *Int. J. Fract.* **100** 121–31
[16] Cherepanov O I., Smolin I Yu, Stefanov Yu P and Makarov P V 1999 *Comput. Mater. Sci.* **16** 25–31
[17] Karakulov V V and Smolin I Yu 2016 *AIP Conf. Proc.* **1760** 020024
[18] Karakulov V V, Smolin I Yu and Skripnyak V A 2013 Numerical procedure of forecasting effective mechanical properties of stochastic composites under shock-wave loading with allowance for the structure evolution *Tomsk State Univ. J. Math. Mech.* **24** 70–7 (in Russian)
[19] Mikushina V A, Smolin I Yu and Sidorenko Yu N 2017 Prediction of mechanical properties of ceramic biocomposite on the basis of numerical modeling *Trans. Tomsk State Univ.* vol 300 *Series Phys.-Math.: Current Issues of Continuum Mechanics and Celestial Mechanics,* ed M Yu Orlov (Tomsk: Tomsk State University Press) pp 107–10 (in Russian)
[20] Sidorenko Yu N 2004 Ph. D thesis (Tomsk: Tomsk State University) p 146 (in Russian)
[21] Wang A S D and Yan K C 2005 *Compos.* A **36** 1335–46
[22] Timoshenko S P and Goodier J N 1970 *Theory of Elasticity* (New York:McGraw-Hill)
[23] Sovetova Yu V, Sidorenko Yu N and Skripnyak V A 2013 *Phys. Mesomech.* **16** 59–65
[24] Segerlind L J 1984 *Applied Finite Element Analysis* (New York: John Wiley)

IOP Conf. Series: Journal of Physics: Conf. Series **919** (2017) 012014 doi:10.1088/1742-6596/919/1/012014

Numerical modeling of the strain of elastic rubber elements

E N Moskvichev[1], A V Porokhin[2] and I V Shcherbakov[2]

[1] Tomsk State University, 36 Lenin Ave., Tomsk 634050, Russia
[2] JSC «Scientific-Production Centre «Polyus», 56 Kirova St., Tomsk 634050, Russia

E-mail: porokhin93@mail.ru

Abstract. A comparative analysis of the results of experimental investigation of mechanical behavior of the rubber sample during biaxial compression testing and numerical simulation results obtained by the finite element method was carried out to determine the correctness of the model applied in the engineering calculations of elastic structural elements made of the rubber. The governing equation represents the five-parameter Mooney-Rivlin model with the constants determined from experimental data. The investigation results showed that these constants reliably describe the mechanical behavior of the material under consideration. The divergence of experimental and numerical results does not exceed 15%.

1. Introduction

An increase in the role of finite element modeling in the engineering calculations is provided by the rapid development of computing techniques and software products. Nowadays, the use of such technologies makes it possible to reduce considerably the time and efforts required to determine the operational parameters of the designed products and to identify design flaws, in particular, the stress-strain state of the products made of hyperelastic materials, and to make the necessary changes at the design stage [1 – 3]. The specific feature of hyperelastic materials is the ability to preserve elastic properties at high strain. Many polymers [4], in particular, the rubber, can serve as an example of such materials whose behavior model under mechanical impact cannot be described by Hooke's law. The reliable description of their mechanical behavior at high strain is the main purpose of the engineering predictions of strength and durability of the important devices and their elements [5].

Currently, many software products are developed to solve such problems. User can achieve the required convergence of the calculation results using a set of various input parameters inherent in a particular material model. The examples of such problem solutions are described in many literature sources, for example, [6 – 9]. Since the hyperelastic models of polymers are very popular, their description and application instructions are well known [6]. However, there is lack of the studies devoted to the description of nonlinear hyperelastic behavior of the selected material class. Therefore, the investigation in this field is very urgent.

In this work, a nontrivial approach with the Mooney-Rivlin model applied as a governing equation is used and it is accompanied by experimental study of mechanical behavior of the rubber sample during biaxial compression testing. The procedure realized by the authors is similar to that in [8] where the stress-strain state of the elastic element of vibration isolator is examined in the NX NASTRAN software environment and the results are compared with the corresponding experimental data.

IOP Conf. Series: Journal of Physics: Conf. Series **919** (2017) 012014 doi:10.1088/1742-6596/919/1/012014

The purpose of this study is to develop the physical and mathematical model and to select the governing equation that reliably describes the behavior of the rubber, which is used in the production of elastic elements of vibration isolators, at strain conditions.

2. Experimental procedure

The standard samples of type *B* are used as an object of research (figure 1, where *1* – rubber cylinder, *2* – steel plates glued to the ends of cylinder). The cylindrical samples have been made by curing the rubber IRP-1378-1 TR 38005924-2002 whose operating temperature under the long exposure ranges from -60 to +50 °C and the hardness is of 40 - 50 Shore units (*A*). The vibration dampers made of such rubber have passed all of the tests being part of the shock absorbers of ARMOO [10] and ARMOO-M series [11]. The plates made of the structural steel have been glued to the ends of cylindrical samples with the glue 88-SA TR 38 1051760–89. For all samples, the time between curing of the elastic element and its testing is not less than 24 hours and more than 1 month.

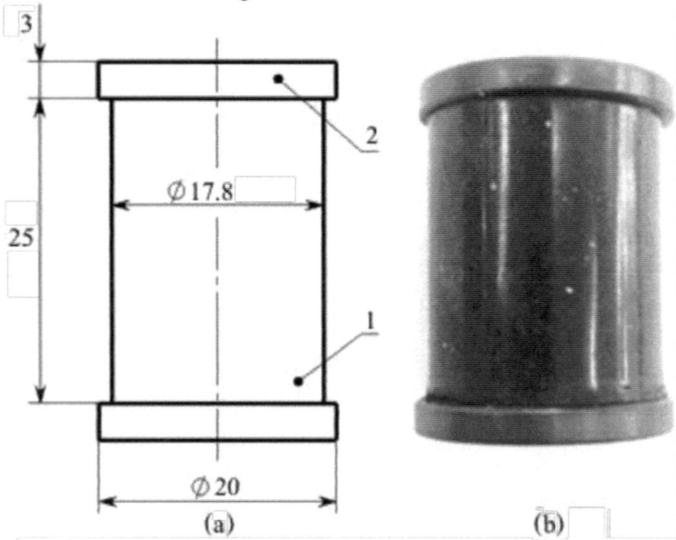

Figure 1. Standard samples of type *B* for determining the rubber properties: scheme (a), real configuration (b).

The tests were carried out using the method *C* according to GOST ISO 7743-2013, which established the methods for determining the mechanical properties of the rubbers and thermoplastic elastomers during compression. The stress-strain curves for the considered material were obtained as a result of biaxial compression of the five prepared samples using the INSTRON 4483 unit equipped with the devices for automatic recording of the force versus strain curve with an accuracy corresponding to 1 class. In the tests, the samples were compressed at a rate of 10 mm / min until the strain of 70% was achieved. The control recording of the force and strain values was implemented after the samples had passed three loading-unloading cycles with an interval of 1 minute between them. When calculating the stress caused by exerted force, the sample cross-sectional area was assumed to be constant and equal to the initial one.

3. Statement of the problem for hyperelastic material modeling

The study of the sample model was carried out using the standard formulation of the problem of continuum mechanics including the equilibrium equation, kinematic correlations, and constitutive equations with an application of the ANSYS software. The Mooney-Rivlin approach [6], which describes the nonlinear elastic behavior of a material at high strain, was taken as a basis of the governing correlations. The strain energy density was assumed to be a function of linear combination of two invariants of the Cauchy-Green strain tensor whose general form can be written as:

$$W = \sum\nolimits_{m=1}^{M} D_m (J-1)^{2m} + \sum\nolimits_{p,q=0}^{N} C_{pq} (\bar{I}_1 - 3)^p (\bar{I}_1 - 3)^q , \tag{1}$$

where W is the strain energy density (strain energy per unit volume); D_m are the constants related to the volumetric response; $J=det(F)$ is the determinant of the gradient $\boldsymbol{F}$ of elastic deformations; C_{pq} are the constants associated with the form change $(C_{00} = 0)$; $\bar{I}_1, \bar{I}_2$ are the first and second invariants of the deviator deformations.

The Mooney-Rivlin model written in the five-parameter form is as follows:

$$W = \frac{1}{d}(J-1)^2 + C_{10}(\bar{I}_1 - 3) + C_{01}(\bar{I}_2 - 3) + C_{11}(\bar{I}_1 - 3)(\bar{I}_1 - 3) + C_{20}(\bar{I}_1 - 3)^2 + C_{02}(\bar{I}_2 - 3)^2 , \tag{2}$$

$$d = \frac{(1-2\nu)}{C_{10} + C_{01}} , \tag{3}$$

where d is the material incompressibility parameter; $C_{10}, C_{01}, C_{11}, C_{20}, C_{02}$ are the Mooney-Rivlin constants; ν is the Poisson's ratio.

The system of equations was solved using the finite element method in Lagrangian coordinates. The biaxial compression of cylindrical sample corresponding to the implemented experiment was simulated. On the sample lower end surface, the boundary conditions prohibiting the linear displacements along the axes were imposed due to the rigid gluing of the end to the steel plate. As for the upper end, the axial displacements u_0 whose values were limited to a maximum strain of 70% were assigned. The linear displacements in the directions perpendicular to the cylinder axis were also forbidden. The general form of the boundary conditions according to the scheme in figure 2a can be presented as follows:

$$u_{x,y,z}|S_2 = 0 , \tag{4}$$

$$u_{x,z}|S_1 = 0; u_y|S_1 = u_0 , \tag{5}$$

$$\sigma|S_3, S_4, S_5 = 0 , \tag{6}$$

where $u_{x,y,z}$ are the displacement vector components defined on the corresponding model surfaces, $\sigma|S_3,S_4,S_5$ are the forming surfaces of cylindrical rubber sample and steel plates.

The finite elements in the form of hexahedrons were used to create the finite-element model of the sample (figure 2b). The grid density was determined from the grid convergence condition.

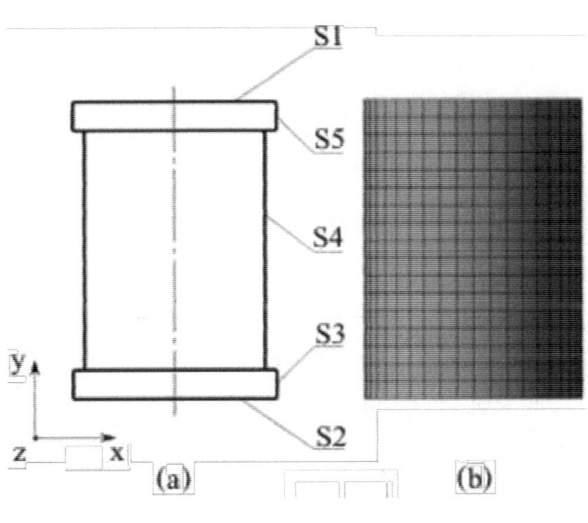

Figure 2. Boundary conditions (a) and finite-element model of the sample (b).

4. Results and discussion

The experimental results of the loading of the five samples show that the strain up to 70% is elastic; the stress-strain curves have an evident nonlinear character (figure 3).

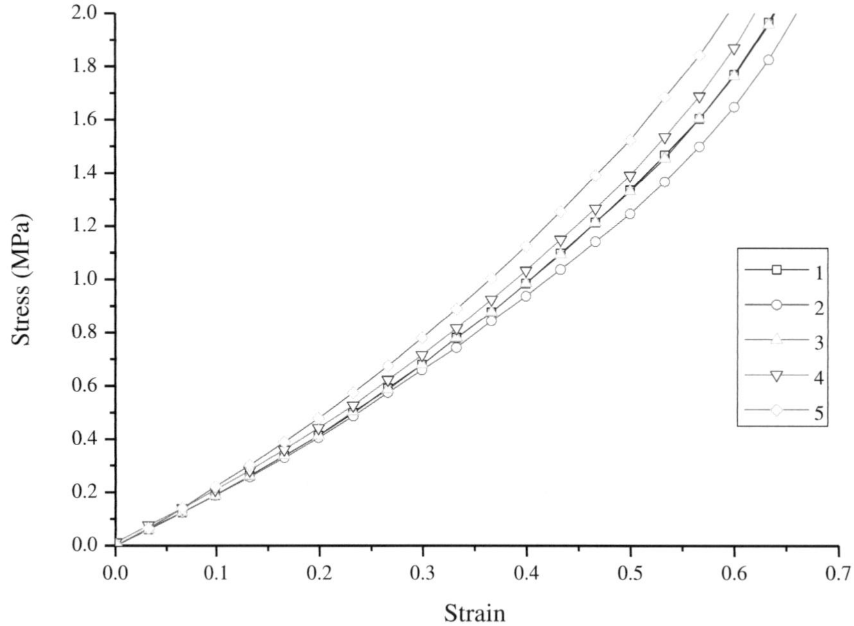

Figure 3. Stress-strain curves obtained using the INSTRON 4483 unit for the five samples.

Based on the values obtained in experiment, the constants for the five-parameter Mooney-Rivlin equation were calculated (Table).

Table. The values of the constants for the Mooney-Rivlin model.

Constant	C_{10}	C_{01}	C_{11}	C_{20}	C_{02}
Value, MPa	1.63	-1.33	-1.22	1.91	0.27

The compression testing of the rubber samples is accompanied by the essential changes in their cross-section area. Since the values of the parameters describing the stress-strain state are represented in true coordinates in the ANSYS software, the evaluation of simulation results when comparing with experimental data requires taking into account the change in the area of experimental sample. Therefore, the experimental curves were converted into true values.

Assuming that the flat cross-section hypothesis is true and the stress distribution along the cross-section of rubber sample under compression is uniform, the true stress values can be determined by the formula [12]

$$\sigma_{tru} = \frac{\sigma_{con}}{1 + 2\nu\varepsilon_1}, \tag{7}$$

where σ_{tru} are the stress values in true coordinates, σ_{con} are the stress values in conditional coordinates, ε_1 is the sample strain.

Based on the experimental data and numerical simulation results, the true stress-strain curves were plotted (figure 4).The confidence intervals denoted on the experimental curve correspond to the permissible statistical error. The maximum discrepancy in the values of confidence interval and simulation results is of 15%.

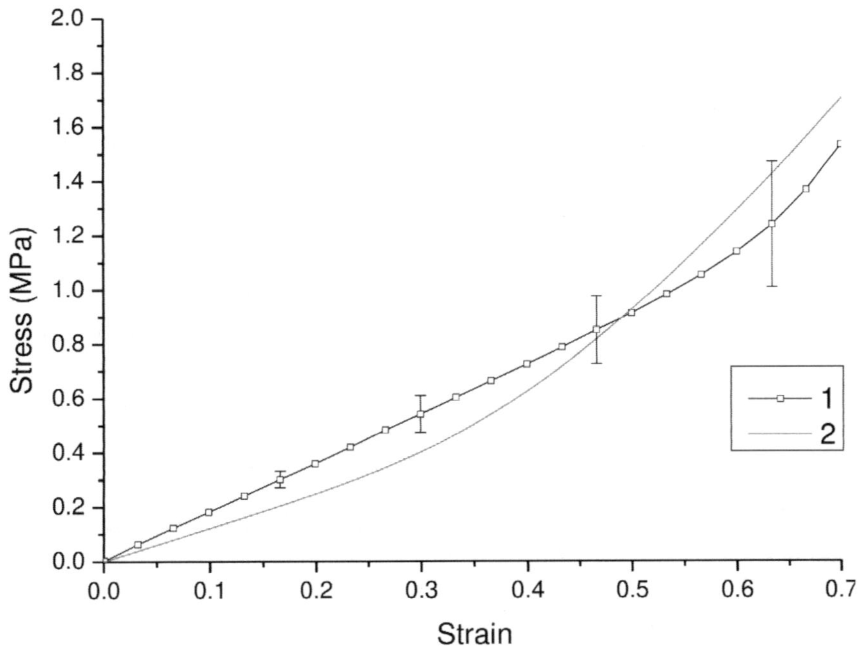

Figure 4. True stress-strain curves obtained experimentally (1) and numerically (2)

A comparison of the finite-element model (figure 5a) and real sample (figure 5b) under biaxial compression confirms that the mathematical model is correct because it accurately describes the experimental sample behavior, and the stress arising in the test sample is approximately similar to that obtained numerically.

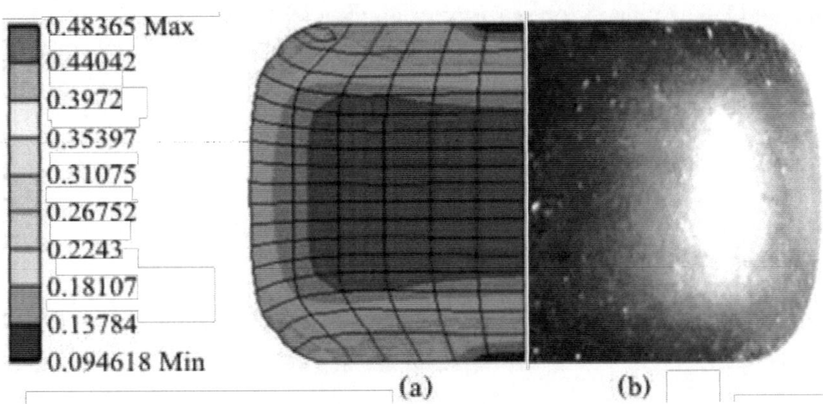

Figure 5. Strain distribution in the sample under biaxial compression at the strain of *48%*: mathematical model (a), experimental sample (b).

5. Conclusion

An experiment on compression of the samples made of the rubber IRP-1378-1 was carried out. It showed that the tested material could strain elastically up to 70%, and the strain exhibited a strong nonlinear character.

The physical and mathematical model developed in the ANSYS software environment describes the nonlinear hyperelastic behavior of the material when compressing the standard cylindrical samples. The five-parameter Mooney-Rivlin model was chosen as the governing equation whose constants were determined from experimental data. The comparison of experimental and numerical results made it possible to determine the maximum discrepancy which is equal to 15%. Consequently, it was stated that the mathematical model accurately described the real strain behavior of nonlinear elastic sample at high strain degree.

The mathematical model presented in this paper can be used in the engineering calculations of elastic elements of vibration isolators to determine the operating parameters at the design stage.

Acknowledgements
The work was supported by the Russian Fund of Fundamental Research (grant No. 16-08-00037).

References

[1] Bodrov A S, Olimpieva N V, Zelepugin A S and Zelepugin S A 2015 Vestnik Tomskogo gosudarstvennogo universiteta. Matematika i mekhanika **5** 37

[2] Sovetova Y V, Sidorenko Y N and Skripnyak V A 2014 Vestnik Tomskogo gosudarstvennogo universiteta. Matematika i mekhanika **2** 77

[3] Gerasimov A V and Pashkov S V 2014 Vestnik Tomskogo gosudarstvennogo universiteta. Matematika i mekhanika **3** 57

[4] Kozulin A A and Skripnyak V A 2005 8th Korea-Russia International Symposium on Science and Technology - Proceedings: KORUS 2004 **3** 29

[5] Skripnyak V A and Kozulin A A 2007 Russian Physics Journal **50** 633

[6] Kuznetsov V V 2016 Energoeffektivnye i resursosberegayushchie tekhnologii i sistemy **2** 75

[7] Murina T A, Kuznetsov V V and Vodakov V N 2016 Energoeffektivnye i resursosberegayushchie tekhnologii i sistemy **4** 50

[8] Balakin P D, Krasotina L V and Krivtsov A V 2016 Omskiy nauchnyy vestnik. Mashinostroenie **3** 5

[9] Kobayashi T et al. 2013 SAE Techinical Paper 53 8

[10] Tselishchev G P. Sbornyy rezinometallicheskiy amortizator s osevym ogranichitelem armoo [Assembler rubber-metal shock absorber with armoo axial limit]. Patent RF, no. 2358167, 2009

[11] Tselishchev G P. Amortizatory rezinometallicheskie s osevym podatlivym ogranichitelem armoo-m [Rubber-metal shock absorbers with axial compliant limiter armoo-m]. Patent RF, no. 2306462, 2007

[12] Ropay V A and Zaichenko V A 2012 Collection of scientific papers **39** 51

IOP Conf. Series: Journal of Physics: Conf. Series **919** (2017) 012015 doi:10.1088/1742-6596/919/1/012015

Modelling stress-strain state in oxide ceramics under thermal shock

V Promakhov, I Zhukov, Ya Dubkova, V Platov, N Grunt and N Shulz

Tomsk State University, 36 Lenina Avenue, Tomsk 634050 Russian Federation
E-mail: vvpromakhov@mail.ru

Abstract. Properties of ZrO_2 -Al_2O_3 ceramic composites were studied in the framework of this research. The porous medium significantly affects the strength of the composite under thermal shock. In the areas near pores, concentrators emerge in a combined stress state with the maximum normal stresses under tension being equal to the compressing ones in absolute values and reaching 2.6 GPa (clusters of pores, 6% porosity) and 2.7 GPa (irregular distribution of pores, 12% porosity). High levels of stress lead to cracks in bridges between pores. The growth of porosity from 6% to 12% increases the number of dangerous zones

1. Introduction

Thermal stability - an important functional property to materials, operating under frequent cyclic temperature changes. Some of the heat-resistant ceramic materials usually referred designed to work in conditions of thermal shock, high temperature gradients, as well as cyclical effects [1, 2, 3]. The questions of theoretical and experimental investigation of the thermal stability of the composite ceramic materials are relevant in the present. The objective of this research is to modeling stress-strain state in oxide ceramics under thermal shock.

2. Description of Model Parameters

The purpose of this work was to use conventional approaches of solid mechanics to evaluate the stress-strain state resulting from temperature distribution within the following oxide ceramic specimens under thermal shock with inclusions and porous structure (figure 1).

We show that it is possible to take into account experimental data on the microstructure of the Al_2O_3 – ZrO_2 nanocomposite specimens at hierarchical scale levels (nano, micro, meso, and macro) when creating model representative volumes of ceramic materials represented in the figure with and without porosity structure. Using the data on representativity of volumes from [4, 5], models were created representing structured representative volumes at mesoscopic level for mathematical simulation with due consideration of distribution of precipitate particles of ZrO in Al_2O_3 matrix in model a); presence of clusters of pores (6% porosity) in model b); and with due account for irregular distribution of pores (12% porosity) and precipitate particles of ZrO_2 (in model c). The dimensions of model specimens are 7.5 × 7.5 μm. Designations (a), (b) and (c) in figures hereinafter correspond to modeled meso-volumes depicted in figure 1.

IOP Conf. Series: Journal of Physics: Conf. Series **919** (2017) 012015 doi:10.1088/1742-6596/919/1/012015

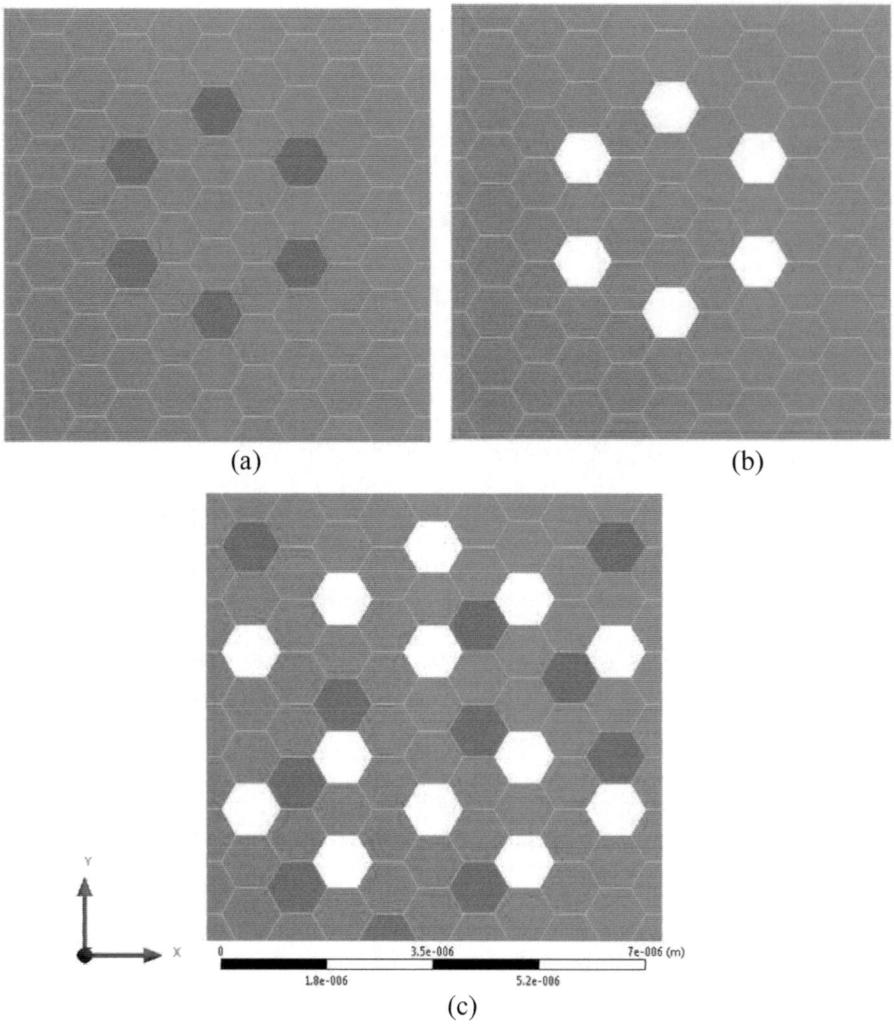

Figure 1. Models of meso-volumes of oxide ceramics based on Al_2O_3 with precipitate particles of ZrO_2 (a); with due consideration of pore clusters (b); with due consideration of irregular distribution of pores and precipitate particles of ZrO_2 (c).

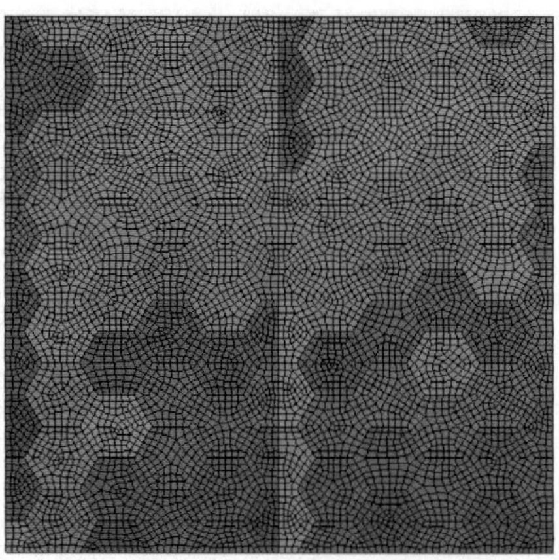

Figure 2. Gridded model.

The figure 2 shows a gridded model. Geometric models of representative meso-volume elements are discretized by gridded models consisting of quadrilateral elements. There are approximately 7,000 elements on the grid. The choice of the grid density and dimensions of finite element was based on the results of grid convergence evaluation. The study also addressed a related thermomechanical problem to model thermal shock in ceramic composite. The problem was solved in two steps. First, we tackled the temperature problem where we had to determine the distribution of temperature fields in the material as a result of rapid cooling. Second, we used the results of the temperature problem to determine the stress-strain state of model specimens resulting from the thermal expansion of the material. The gridded model remained the same for both types of problems. The problem was solved as a two-dimensional one, i.e. the thickness of the model is much smaller than the other two dimensions. Such a statement allows analyzing the distribution of temperature and stress in the two-dimensional approximation.The composite models under consideration have a grain structure and are similar in outside measurements, which can be idealized by the perimeter of computational domain ABCD shown on the scheme (figure 1). In the temperature problem, the computational domains are subjected to intense cooling by the side AB. We chose aluminum oxide Al_2O_3, as the matrix material and ZrO2 as the inclusion material.

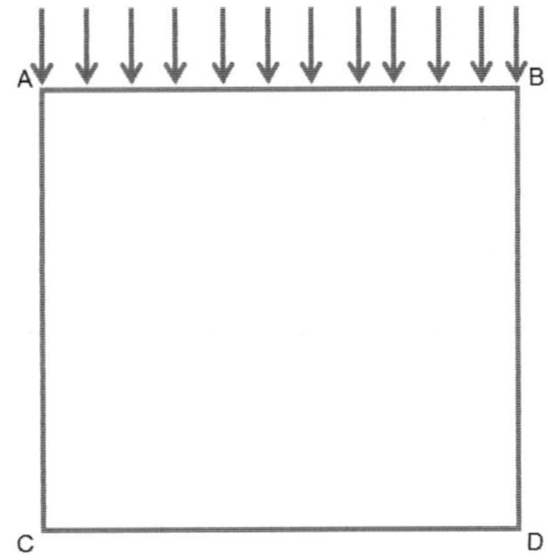

Figure 3. Schematic representation of computational domain.

In solving the temperature distribution problem in the computational domain, the system of linear algebraic equations corresponding to the grid comprised approx. 7,000 equations. The nonstationary heat conduction problem was solved by an implicit scheme [6].

For calculation we used the thermophysical material properties shown in Table 1.

Table 1. Thermophysical material properties

	Al_2O_3	ZrO_2
[a] C, J/(kg °C),	1106	660
ρ, kg/m^3	3800	6000
[b] K_{xx}, K_{yy}, W/(m °C)	30	3

[a] C is heat capacity, ρ is density.
[b] K_{xx}, K_{yy} are coefficients of heat conductivity along x and y axes.

3. Solution to the Heat Conductivity Problem

For the temperature field in the two-layer plate to be found by the finite element method [7, 8], a nonstationary heat conduction problem is solved for the domain ABCD. At the interface of the material differing in physical and mechanical characteristics, the thermal contact is perfect: the temperatures and heat fluxes are equal [7]. The edges AC and BD are assigned the symmetry conditions:

$$\frac{\partial T}{\partial x}|_{AC}=0,\ \frac{\partial T}{\partial x}|_{BD}=0. \tag{1}$$

These conditions (called sometimes heat insulation conditions) mean that the introduced boundaries artificially narrow down the region under study and in a real situation the adjacent region has the same characteristics as the computational domain.

The edges AB and CD kept respectively at temperatures of 22 °C and 800 °C are assigned the Dirichlet conditions:

$$T|_{AB}=22\ °C,\ T|_{CD}=800\ °C. \tag{2}$$

At the point in time t = 0, the temperature over the entire domain ABCD is 800 °C:

$$T(x,y)=800\ °C \tag{3}$$

In solving the problem by the finite element method, the heat conduction equation with boundary conditions (1) and (2) is reduced to minimization of the functional [6]:

$$\chi=\int_V 0.5\left[K_{xx}\left(\frac{\partial T}{\partial x}\right)^2+K_{yy}\left(\frac{\partial T}{\partial y}\right)^2+2\lambda\frac{\partial T}{\partial t}T\right]dV, \tag{4}$$

where $\lambda = c\rho$, K_{xx}, K_{yy} are the thermal conductivities; c is specific heat capacity; ρ is the density, and $V$is the domain volume.

The temperature field distributions in the material of the specimens are presented in figure 4.

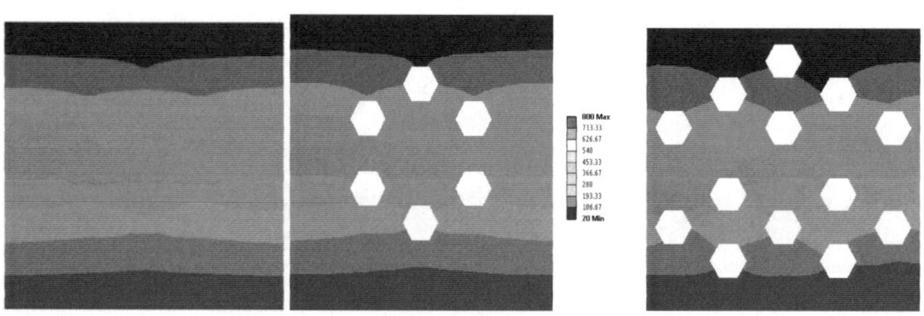

(a) (b) (c)

Figure 4. Temperature field distributions in model specimens.

The distribution of temperature fields in computed areas is nonuniform due to different thermophysical characteristics of material of matrix and inclusions, presence of voids.

4. Analysis of Stress-Strain State

The next step makes use of the results from the temperature problem. As it has been established, intense heating of the heterogeneous plate gives rise to thermal stresses. The thermal stresses are estimated by solving the stress-strain state problem. It is assumed that at the contact boundary of the grains, the adhesion is perfect: the displacements and stresses are equal. The stress-strain state problem is solved using the finite element method (FEM) [6]. In terms of strength estimation, if there is a temperature with large gradient under thermal shock, this indicates a dangerous state.

The potential energy of an elastic solid acted by surface forces P_x, P_y, *and* P_z with initial strain ε_0 is [5]:

$$\Pi = \sum_e^n \int_{V^{(e)}} \frac{1}{2}\{U\}^T[B^{(e)}]^T[D^{(e)}][B^{(e)}]\{U\}dV - \sum_e^n \int_{V^{(e)}} \frac{1}{2}\{U\}^T[B^{(e)}]^T[D^{(e)}][B^{(e)}]\{\varepsilon_0^{(e)}\}dV$$
$$-\sum_e^n \int_{S^{(e)}} \frac{1}{2}\{U\}^T[N^{(e)}]^T \begin{Bmatrix} P_x^e \\ P_y^e \\ P_z^e \end{Bmatrix} dS, \qquad (5)$$

where $\{U\}$ is the displacement vector; $[B^{(e)}]$ is the strain-displacement matrix containing derivatives of shape function; $[D^{(e)}]$ is the elasticity matrix; $\{\varepsilon_0\}$ is the initial strain vector; $[N^{(e)}]$ is the matrix containing shape functions; P_x, P_y, P_z are the surface load intensities along the axes; $S^{(e)}$ is the area of elements acted by the surface load; V and S are the volume and area of the elastic solid.

Resulting equations of the finite element method are derived from minimization conditions of the total potential energy functional Π of the mechanical system:

$$\frac{\partial \Pi}{\partial \{U\}} = 0 \qquad (6)$$

The minimization of the functional Π gives the system of linear algebraic equations:

$$[K]\{U\} = \{F\} \qquad (7)$$

,

where $[K]$ is the global stiffness matrix of the mechanical system and $\{F\}$ is the force vector.

The global stiffness matrix of the finite element grid is equal to the sum of stiffness matrices of individual elements. The global force vector $\{F\}$ is equal to the sum of force vectors of individual elements $\{f^{(e)}\}$:

$$\{F\} = \sum_e^n \{f^{(e)}\} \qquad (8)$$

The force vector of a single finite element with surface forces P_x, P_y, P_z and initial strains is written in the form:

$$\{f^{(e)}\} = \int_{V^{(e)}} [B^{(e)}]^T[D^{(e)}]\{\varepsilon_0^{(e)}\}dV + \int_{S^{(e)}} [N^{(e)}]^T \begin{Bmatrix} P_x^{(e)} \\ P_y^{(e)} \\ P_z^{(e)} \end{Bmatrix} dS \qquad (9)$$

The initial strain vector for the plane stress state is equal to

$$\{\varepsilon_0\} = \alpha \Delta T \begin{Bmatrix} 1 \\ 1 \\ 0 \end{Bmatrix} \qquad (10)$$

where α is thermal expansion coefficient (1/°C) ; ΔT is the temperature difference of a single finite element (°C).

For the plane stress state, the integral related to thermal expansion (initial strain) is equal to

$$\int_{V^{(e)}} [B^{(e)}]^T [D^{(e)}]\{\varepsilon_0^{(e)}\}dV = \frac{Ea_t\alpha\Delta T}{2(1-\nu)}\begin{Bmatrix} b_i \\ c_i \\ b_j \\ c_j \\ b_k \\ c_k \end{Bmatrix}, b_i = y_j - y_k,\ c_i = x_k - x_j \quad (11)$$

where E is Young's module (MPA); ν is Poisson's ratio; a_t is the thickness of an element. The coefficients b_j, c_j, b_k, c_k are derived by circular substitution.

For the plane stress state, the initial strain vector $\{\varepsilon_0\}$ and the elasticity matrix $[D^{(e)}]$ change [6]. The integral related to thermal expansion for plane strain is equal to

$$\int_{V^{(e)}} [B^{(e)}]^T [D^{(e)}]\{\varepsilon_0^{(e)}\}dV = \frac{Ea_t\alpha\Delta T}{2(1-2\nu)}\begin{Bmatrix} b_i \\ c_i \\ b_j \\ c_j \\ b_k \\ c_k \end{Bmatrix} \quad (12)$$

Like in the previous calculation, the grid contained about 7,000 finite elements; the number of algebraic equations whose solutions give the displacement vector was 7,000.

The result of problem solving was displacement, strain and stress fields in the cross-section of the plate ABCD. In solving the plane elastic problem, symmetry conditions were specified in the domain ABCD. As with the thermophysical problem, the introduced boundaries artificially narrow down the dimensions of the area to be analyzed, while in a real case the adjacent area is assumed to have the same characteristics as the calculated one.

At the edge CD, the displacement V along the axis y is equal to zero, i.e.

$$V\,|_{CD} = 0 \quad (13)$$

on the lines AC and BD, the displacement U along the axis x is equal to zero, i.e.

$$U\,|_{AC} = 0\,,\ U\,|_{BD} = 0\,. \quad (14)$$

The stress-strain state was calculated with the physical and mechanical characteristics of the materials and precipitate inclusions presented in the table.

Table 2. Physical and mechanical characteristics of materials.

	Al_2O_3	ZrO_2
E, GPa	380	210
G, MPa	153	84
ν,	0.35	0.35
α, 10^{-6} 1/°C	9.0	0.11

Figure 5 and 6 show surfaces illustrating the distributions of normal stresses σ_y of the stress tensor and σ_{11} of the tensor of main stresses, caused by nonuniform distribution of temperature along the plate thickness, depicted in Fig. 4.

The difference between thermophysical and physicomechanical characteristics of the material of inclusions and matrix as well as porosity and the temperature gradient together affect the stress-strain state of the composite.

The pore-free medium with inclusions suffers the occurrence of negligible tensile strains in the area of the inclusions equal to 588 MPa. The normal stress in the compression area is 298 MPa. Calculations demonstrate that tensile stresses introduce the maximum contribution to stress-strain state of the considered non-porous composite with the clustered distribution of inclusions. Under such conditions of temperature distribution in the considered meso-volume, the composite sustains thermal shock without fracture.

The porous medium appreciably affects the strength of the composite under thermal shock. The consideration of porous medium results in a combined stress state. In the areas near pores, concentrators emerge with maximum normal stresses that under tension are equal to compressing ones in absolute values and reach 2.6 GPa (clusters of pores, 6% porosity) and 2.7 GPa (irregular distribution of pores, 12% porosity). High levels of stress lead to cracks in *bridges* between pores. Increased porosity from 6% to 12% lead to an increase in the number of dangerous zones.

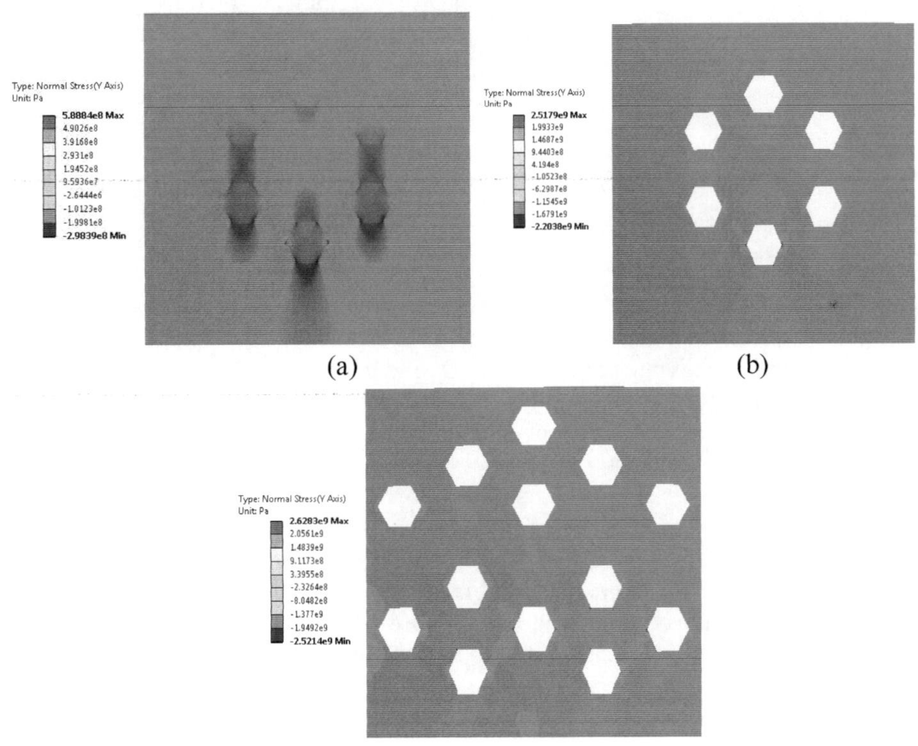

Figure 5. Distribution of normal temperature stresses σ_y in model volumes.

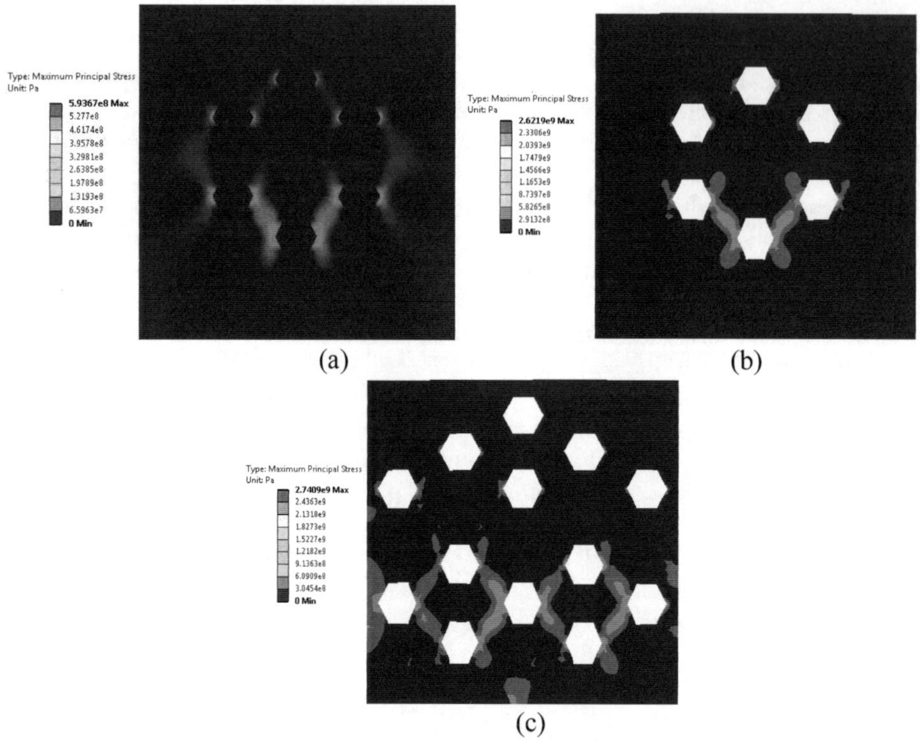

Figure 6. Distribution of temperature stresses σ_{11} in model volumes.

5. Conclusion

The following major factors were revealed as a result of solution of coupled thermomechanical problems on stress-strain state under thermal shock in representative volumes of ceramic composites having Al_2O_3 matrix with precipitate particles of ZrO_2 and pores:

- Distribution of temperature in material is nonuniform; the distribution of temperature fields is appreciably distorted by inclusions and pores with thermophysical characteristics that differ from those of matrix material;

- The pore-free medium with inclusions suffers the occurrence of negligible tensile strains (less than 600 MPa) in the area of the inclusions. The composite sustains thermal shock without fracture.

- The porous medium significantly affects the strength of the composite under thermal shock; in the areas near pores, concentrators emerge in a combined stress state with the maximum normal stresses under tension being equal to the compressing ones in absolute values and reaching 2.6 GPa (clusters of pores, 6% porosity) and 2.7 GPa (irregular distribution of pores, 12% porosity). High levels of stress lead to cracks in *bridges* between pores. The growth of porosity from 6% to 12% increases the number of dangerous zones.

Acknowledgments

The work was financially supported from Grant of the President of the Russian Federation 2424.2017.8 agreement № 14.Y30.17.2424-MK.

References

IOP Conf. Series: Journal of Physics: Conf. Series **919** (2017) 012015 doi:10.1088/1742-6596/919/1/012015

[1] Korobenkov M V and Levkov R V 2016 The dynamic resistance of the graded ceramic materials. *Proc. Int. Youth Scientific Conf. "Current issues of continuum mechanics and celestial mechanics"* vol 57, ed M Yu Orlov (Tomsk: Tomsk State University Press) p 57.

[2] Skrupnyak E G et all 2010 Modeling of the mechanical behavior of ceramic composites with a transformation-hardened matrix under dynamic influences. Tomsk State University Journal of Mathematics and Mechanics **2**.

[3] Promakhov, V. V., Zhukov, I. A., Vorozhtsov, S. A., Zhukov, A. S., Vorozhtsov, A. B. (2016). Thermal Shock-Resistant Ceramic Composites Based on Zirconium Dioxide1. Refractories and Industrial Ceramics, 56(6), 610-614.

[4] Lyukshin P A, Lyukshin B A, Matolygina N Yu and Panin S V 2011 Modeling stress-strain state and stability loss of thermal-barrier coating under thermal shock. Physical mesomechanics **14** **1** 33–41.

[5] Skripnyak V V, Skripnyak E G, Vaganova I K and Skripnyak V A 2013 Multiscale modeling of deformation and fracture of structured solid bodies. Problem of determination of representative volume for dynamic loading conditions. Russian Physics Journal. **56**. No. 7-3. 80-82.

[6] Zenkevich O and Morgan K 1986 *Finite Elements and Approximation* (Moscow: Mir).

[7] Segerlind L 1979 *Application of Finite Element Method* (Moscow: Mir).

[8] Plyatt Sh N 1974 *Estimation of Temperature Fields of Concrete Constructions* (Moscow: Energia).

IOP Conf. Series: Journal of Physics: Conf. Series **919** (2017) 012016 doi:10.1088/1742-6596/919/1/012016

Numerical simulation of single-stage axial fan operation under dusty flow conditions

L L Minkov and E V Pikushchak

National Research Tomsk State University, 36, Lenin av., Tomsk, 634050, Russia

E-mail: lminkov@ftf.tsu.ru, liza@ftf.tsu.ru

Abstract: Assessment of the aerodynamic efficiency of the single-stage axial flow fan under dusty flow conditions based on a numerical simulation using the computational package Ansys-Fluent is proposed. The influence of dust volume fraction on the dependences of the air volume flow rate and the pressure drop on the rotational speed of rotor is demonstrated. Matching functions for formulas describing a pressure drop and volume flow rate in dependence on the rotor speed and dust content are obtained by numerical simulation for the single-stage axial fan. It is shown that the aerodynamic efficiency of the single-stage axial flow fan decreases exponentially with increasing volume content of dust in the air.

1. Introduction

It is known that the standard practice for an axial fan characterization are the engineering techniques with the following using of the available experimental database for obtaining the correction coefficients allowing one to precise the specific fans characteristics [1-3].

As a rule, the creation of such an experimental database for new type of fans requires the carrying out a series of experiments that is a resource and wasteful action.

Availability of the present-day computational software [4] allows one to carry out the direct numerical modeling of fans operation and to predict adequately its characteristics that leads to the essential reduction of material costs [5].

One of important characteristics of an axial fan is the aerodynamic efficiency representing a ratio of the air flow power produced by the fan to the power of the moment of forces acting on blades of the fan wheel (rotor). The investigation of the influence of rotational speed of rotor and angle of blades position in relation to a fan hub for dusty-free air on the aerodynamic efficiency of the axial fan was carried out in [6].

Wide use of axial fans for ventilation of coal mines excavations requires the research of influence of dust content in air on the main characteristics of such fans.

In this paper the research of influence of coal dust content in air on the aerodynamic efficiency of the single-stage axial fan with the straightening device is carried out with using of numerical modeling on the basis of Ansys-Fluent Software.

2. Physical model of the axial fan

The axial fan, fig. 1 consisting of the rotor 1, the straightening device (SD) 2, the directing hub of rotor 3, the directing blade SD 4 and the casing 5 has been considered. The directing blades of the rotor and SD have the identical radius R_1. The external shroud is a pipe with radius R_2. The blades (in quantity N_r) are disposed on the rotor with a constant step and their height is equal to $R_2 - R_1$. The angle between a circumferential direction and a chord of rotor blade profile is equal to α. Length of a rotor blade chord is equal to b_r. The directing blades (in quantity N_s) with their profile chords (length b_s) are parallel to the directing blade of SD are disposed on the SD with a constant step. Shapes of rotor blades and SD are given in the paper [6]. Rotor blades rotate at angular speed ω, providing the air movement in the direction parallel to the generatrix of hub and increasing the total pressure of air [3, 7, 8].

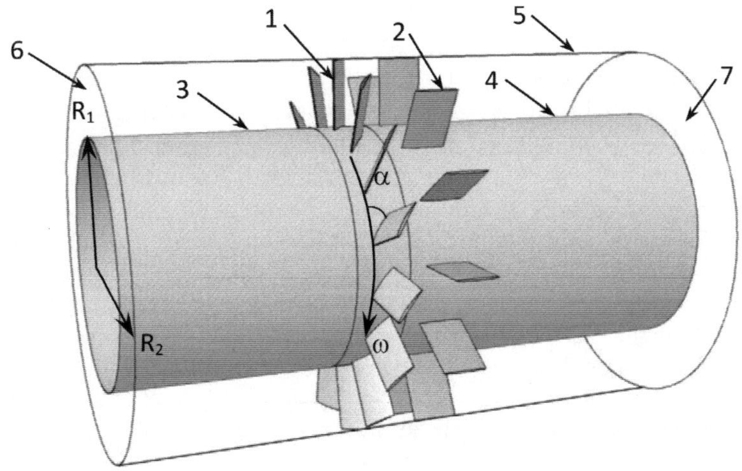

Figure 1. Axial fan

As the fan studied corresponds to the mine fan of local ventilation then, because of low values of Mach number, we neglect the effect of compressibility, and we suppose that a flow is isothermal.

3. Mathematical model of the axial fan

As the rotor and SD free for air mass have periodically repeating parts (angles of periodicity are $2\pi/N_r$ and $2\pi/N_s$ for rotor and SD respectively), then we research the flow in the domain represented in the fig. 2.

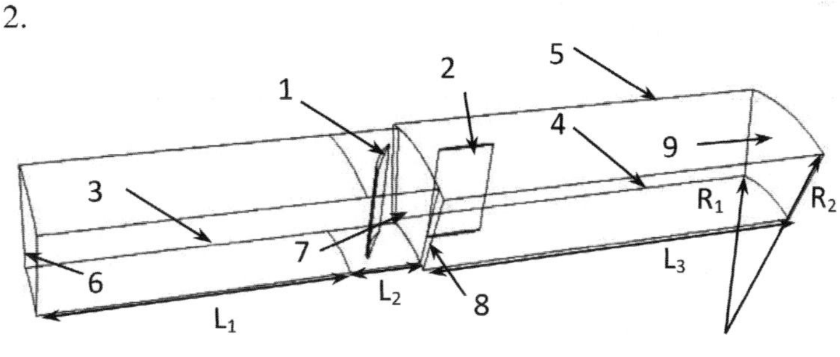

Figure 2. Geometry of domain
1 – rotor blade; 2 – SD blade; 3 – directing rotor hub; 4 –directing SD hub; 5 – shroud; 6 – entrance surface of a rotor; 7 – output surface of a rotor; 8 – entrance surface of SD; 9 – output surface of SD.

The approach of the interpenetrating continua was applied for modeling the dusty air flow via the axial fan. The system of the governing equations written down in the rotating frame of reference consists of the mass conservation equations for air and dust, the momentum conservation equations for air and dust, the equation of turbulent kinetic energy for mixture and the equation of dissipation rate of turbulent kinetic energy for mixture.

Boundary conditions for the system of governing equations were set as follows. At the inlet in the domain the volume fraction of dust particles was set and the total atmospheric pressure of air was specified. At the outlet from the domain the atmospheric static pressure of air was specified. On walls of the shroud, rotor and SD no-slip conditions were set. Turbulent parameters of the flow on walls and in near wall region were defined according to classical k-ε model of turbulence [4].

4. Solution method of the system of governing equations

Ansys Fluent Software was used for solving the problem. The system of the governing equations was solved numerically with use of the Patankar method. Convective terms of the equations were approximated by means of the upwind scheme of the second-order accuracy. Velocity-pressure coupling was carried out using the algorithm SIMPLE on the staggered grid.

For SD the system of the equations was solved in the mfixed coordinate system (ω=0), and for a rotor – in the rotating coordinate system with an angular speed ω relatively longitudinal axis. Recalculation of parameters from an output surface of a rotor on an inlet surface of SD was done on the basis of the concept "mixing plain interface" realized in Ansys-Fluent. According to this concept the profiles of average parameters of total pressure, the directing angle cosines of velocity vectors in the radial, tangential and axial directions, turbulent kinetic energy and turbulent dissipation rate are transferred from the area of the rotor to the SD area which then are used as entrance boundary conditions for the SD area.

The calculation domain was meshed with 340 thousand hexahedral cells. Calculation accuracy was controlled by balance of mass flows. The grid was refined near blades of the rotor and the SD.

5. Results of numerical modeling of the axial fan operation

Calculations of axial fan parameters were carried out under the following parameter values: $\mu = 1.8\cdot 10^{-5}$ Pa·sec; $\rho_{air} = 1.225$ kg/m^3; $\rho_{coal} = 1600$ kg/m^3; $R_1 = 225$ mm, $R_2 = 350$ mm, $b_r = 67$ mm; $b_s = 80$ mm; $L_1 = 400$ mm; $L_2 = 100$ mm; $L_3 = 600$ mm; $N_r = 15$, $N_s = 12$. Rotational speed of the rotor was varied from 500 rpm to 5000 rpm, the dust volume fraction β was varied from 10^{-5} to 10^{-3}. The size of dust particles was equal 10 microns. The value of an orientation angle of a working rotor blade α was equal 30°. As it was shown earlier [6], in case of such a value α the maximum of fan aerodynamic efficiency for dust-free air is reached.

The results of calculation of static pressure drop between input and output of the fan depending on rotational speed of the rotor and the volume content of dust in air are shown in fig. 3.

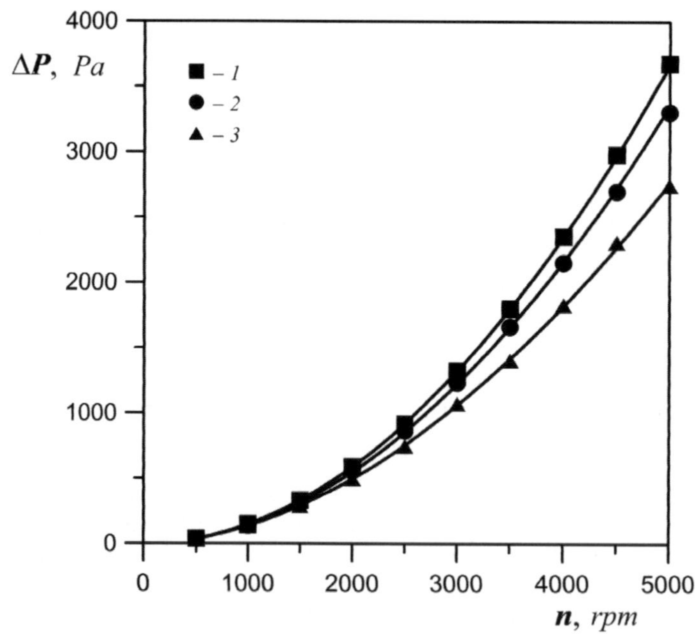

Figure 3. Dependence of the air volume flow rate on a rotor's rotation frequency
1 – $\beta=10^{-5}$, *2* – $5\cdot 10^{-4}$, *3* – 10^{-3}.

It is clear that the increase of rotational speed of the rotor and the decrease of dust content leads to the growth of static pressure drop. Approximation of calculation results using the least-squares method shows that the pressure drop is proportional to a square of a rotor's rotation frequency, i.e. $\Delta P \sim n^2$.

The increase of a rotor's rotation frequency leads to the growth of the volume flow rate, $Q \sim n$, fig. 4. Results of numerical modeling for dependence of the air volume flow rate on a rotor's rotation frequency and a volume content of dust can be fitted by the correlation $Q = A \cdot n \cdot \exp\left[-B(n)\beta\right]$ with the mean error, which is not exceed 2%. In the given formula the flow rate Q has dimension m³/sec, and rotation frequency – rpm, $B(n) = 59.345 \cdot \ln(n) + 68.744$, $A = 3.43 \cdot 10^{-3}$.

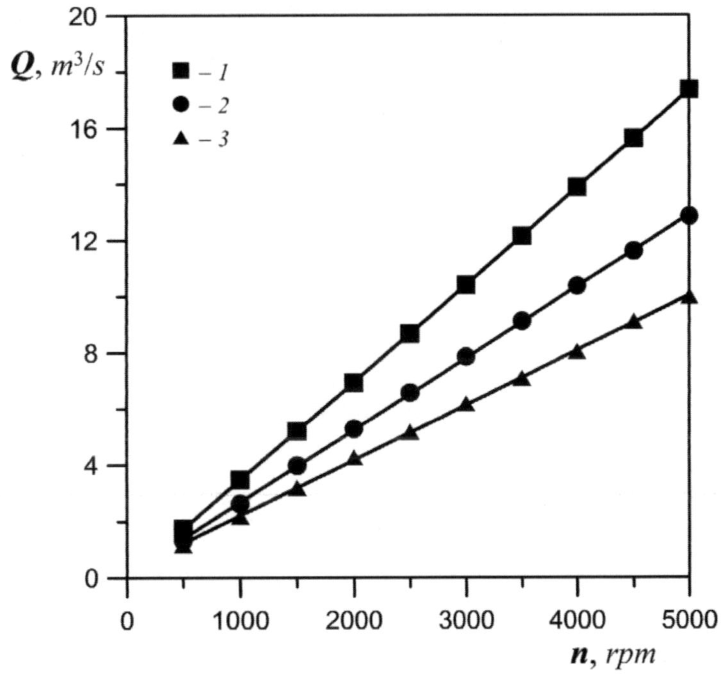

Figure 4. Dependence of pressure drop on rotational speed of rotor
1 – β=10^{-5}, *2* – 5·10^{-4}, *3* – 10^{-3}.

One of the important characteristics of the fan is the dependence of pressure drop on the air volume flow rate. From fig. 5 we can see that the increase of dust content reduces the air volume flow rate under the given pressure drop. All set of calculated data are fitted by a formula close to that, which results from Bernoulli's equation: $\Delta P = \frac{\rho}{2S^2} Q^2 \cdot (1 + 1303.34 \cdot \beta)$, where ρ – density of air, and S – the square of fan cross section.

Another important characteristic of the fan is the power of flow created by a rotor. Its value can be defined as a product of the air volume flow rate and pressure drop: $N_p = Q \cdot \Delta P$. It follows from fig. 3-5 that the power of air flow increases with the growth of the rotational speed of rotor and with reduction of dust content in air. When the rotor rotates, the dust-air flow exerts pressure on the working blades and the moment of forces relatively to longitudinal axis arises. The power of resistance forces, at which the fan has aerodynamic characteristics required, is defined as [6] $N_v = M_1 N_r 2\pi n/60$.

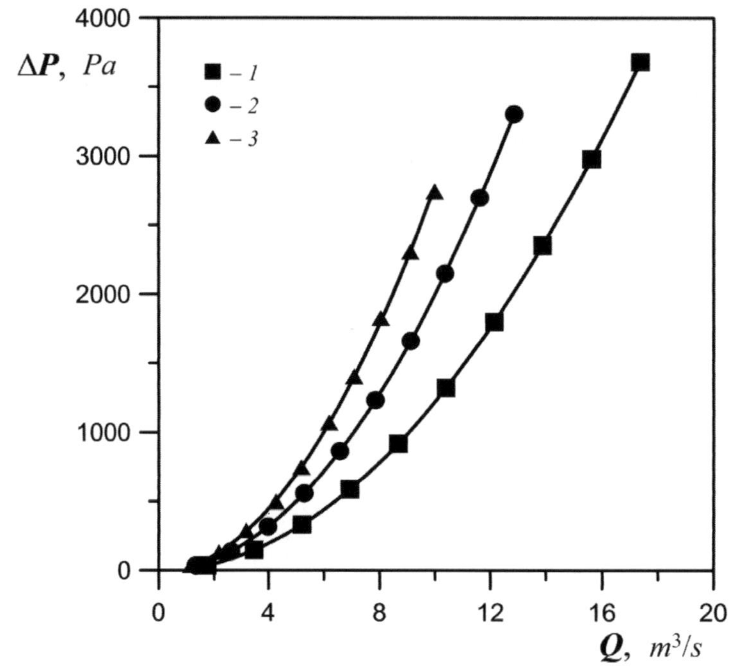

Figure 5. Dependence of pressure drop created in the fan on the air volume flow rate *1* – $\beta=10^{-5}$, *2* – $5\cdot10^{-4}$, *3* – 10^{-3}.

Aerodynamic efficiency of the fan can be defined as the ratio of the power of air flow created by the fan to the power of fan rotor resistance forces, $\eta = \dfrac{N_p}{N_v}100\%$.

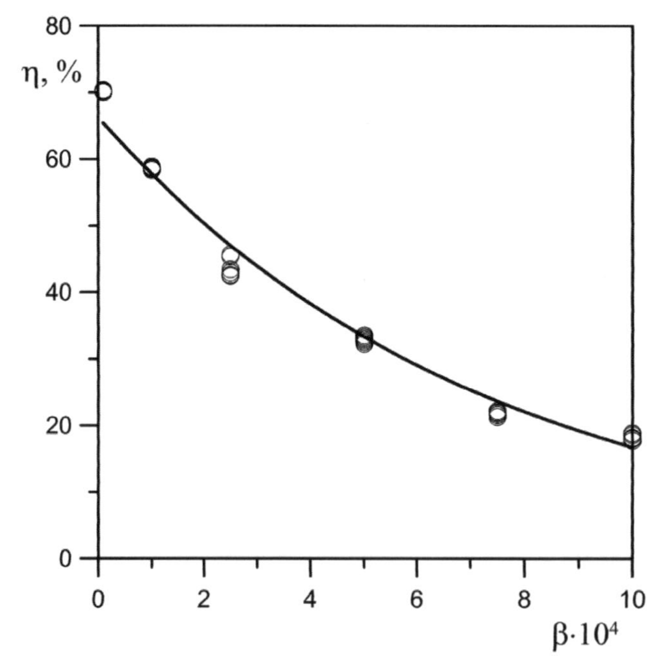

Figure 6. Dependence of fan efficiency on the volume content of dust.

Calculation results of the efficiency expressed in % are given in fig. 6. It is clearly that with the growth of the dust content from 10^{-5} to 10^{-3} the aerodynamic efficiency of the fan exponentially decreases from 70% to 20%. At the same time it is necessary to notice that for rotational speed of a rotor higher than 1000 rpm the efficiency of the fan practically remains constant under determined volume dust particles fraction. The value of the aerodynamic efficiency expressed in percentage can be described by a formula $\eta = A \cdot \exp(-B \cdot \beta)$, where $A = 66.29$, $B = 1347.6$ (a continuous curve on fig. 6).

6. Conclusions

On the basis of numerical simulation with Ansys-Fluent software the dependence of single-stage axial fan aerodynamic efficiency on the volume fraction of coal dust in air varied from 10^{-5} to 10^{-3} is derived.

The dependence of the pressure drop along the fan on the volume air flow rate with taking account of the dust content is obtained.

The formula for the volume dust-air flow rate vs the rotational speed and the volume fraction of coal dust content fits numerical results with the mean error, which is not exceed 2%.

7. Acknowledgements

The reported study was funded by RFBR according to the research project № 15-38-50843 “mol_nr”.

References

[1] Brussilovskiy I V 1986 *Aerodinamicheskiy raschet osevykh ventilyatorov* (Moscow: Mashinostroenie) p 288

[2] Popov N A and Popov N N 2002 K voprosu rascheta aerodinamicheskoy kharakteristiki shakhtnogo osevogo ventilyatora s povorotnymi na khodu lopatkami rabochego kolesa *Gornyi informatsionno-analiticheskiy bulleten* **2** 211–15

[3] Sherstyuk A N 1972 *Nasosy, ventilyatory, kompressory. Uchebnoe posobie dlya vtuzov* (Moscow: Vysshaya shkola) p 344

[4] ANSYS FLUENT Tutorial Guide: Release 14.0. 2011. ANSYS Inc.

[5] Gurina E I 2010 Modeling of the operation of a mine counter rotation fan by means of the fluent suite *Journal of Engineering Physics and Thermophysics* **83** 985–90

[6] Pikushchak E V and Minkov L L 2016 Method of calculation of aerodynamic efficiency of the axial fan *Journal of Mathematics and Mechanics* **1** 90–101

[7] Smirnov I N 1969 *Gidravlicheskie turbiny i nasosy* (Moscow: Vysshaya shkola) p 400

[8] Krivchenko G I 1978 *Gidravlicheskie mashiny: Turbiny i nasosy. Uchebnik dlya vuzov* (Moscow: Energiya) p 320

IOP Conf. Series: Journal of Physics: Conf. Series **919** (2017) 012017 doi:10.1088/1742-6596/919/1/012017

Research of the rotation effect upon the hydrodynamics and heat and mass transport in a chemical reactor

Natalia I Gicheva

National Research Tomsk State University, Faculty of Physics and Engineering, Department of Applied Aerodynamics, 36 Lenin Ave., Tomsk, 634050, Russia

E-mail: tashka-zenit@mail.ru

Abstract. The subject of this research is a chemical reactor for producing tungsten. A physical and mathematical model of fluid motion and heat and mass transfer in a vortex chamber of the chemical reactor under forced and free convection has been described and simulated using two methods. The numerical simulation was carried out in «vortex – stream function» and «velocity – pressure» variables. The velocity field, the mass and the temperature distributions in the reactor were obtained. The influence of a rotation effect upon the hydrodynamics and heat and mass transport was showed. The rotation is important for more uniform distribution of temperature and matter in the vortex chamber. Parametric studies on effects of the Reynolds, Prandtl and Rossbi criteria on the flow characteristics were also performed. Reliability of the calculations was verified by comparing the results obtained by the methods mentioned above. Also, the created model was applied for numerically solving of the classical test problem of the velocity distribution in an annular channel and that of a rotating infinite disk in a stationary liquid. The study findings showed a good agreement with the exact solutions.

1. Introduction

The industry development requires improving the quality of materials. This determines the interest of science to refractory metals having very high durability. One of the most demanded refractory metals in different branches of industry is tungsten. However, the tungsten production and making the products desired shape are very difficult. Therefore necessity of development of new ways of tungsten isolation from its compounds exists. One such promising method is chemical vapor deposition from tungsten fluorides and chlorides. The stoichiometry of tungsten deposition from the mixture of tungsten hexafluoride and hydrogen is

$$WF6 + 3H2 \rightarrow W + 6HF.$$

This reaction can start at a temperature of about 300° C and allow to obtain any thickness of tungsten blanket.

2. The reactor operation scheme

Modeling of viscous gas dynamics and heat transfer is considered in a vortex chamber that present a cylindrical chamber (see figure 1).

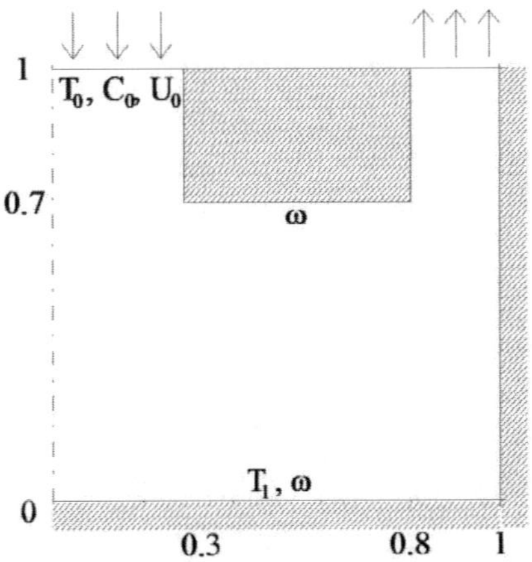

Figure 1. The computational domain of the vortex chamber.

The carrier gas flow having axial velocity U_0, temperature T_0 and concentration C_0 enters from the pipe along the axis above, flows over the rotating disk and exits through the annular channel at the periphery of the top of the vortex chamber. The top and bottom walls of the apparatus are rotating with an angular velocity ω. The bottom wall is maintained at a temperature T_1, and other walls are considered as heat-insulated.

3. Mathematical model

A mathematical model of the processes being considered consists of partial differential equations which includes the fluid dynamics as well as heat and mass transport phenomena along with boundary conditions, incorporating all chemical and physical processes.

3.1. The equations for flow in a vortex chamber

3.1.1. System of equations in «velocity – pressure» variables. By virtue of relatively small velocities and temperature differences, gas is assumed to be incompressible and density change is taken into account only in terms that includes the force of gravity and besides density difference is proportional to the temperature difference with the opposite sign according to the Boussinesq approximation. From above, non-dimensional transport equations of momentum, heat and mass and continuity equation in cylindrical coordinates in the assumption of axial symmetry in angular direction in «velocity – pressure» variables are follows:

$$\frac{\partial u_r}{\partial t}+\frac{\partial (u_r)^2}{\partial r}+\frac{\partial (u_z u_r)}{\partial z}=-\frac{\partial p}{\partial r}+\frac{1}{\mathrm{Re}}\left(\frac{\partial^2 u_r}{\partial r^2}+\frac{1}{r}\frac{\partial u_r}{\partial r}+\frac{\partial^2 u_r}{\partial z^2}-\frac{u_r}{r^2}\right)+\frac{u_\varphi^{\ 2}}{r}-\frac{u_r^{\ 2}}{r}; \quad (1)$$

$$\frac{\partial u_z}{\partial t}+\frac{\partial (u_r u_z)}{\partial r}+\frac{\partial (u_z)^2}{\partial z}=-\frac{\partial p}{\partial z}+\frac{1}{\mathrm{Re}}\left(\frac{\partial^2 u_z}{\partial r^2}+\frac{1}{r}\frac{\partial u_z}{\partial r}+\frac{\partial^2 u_z}{\partial z^2}\right)-\frac{u_r u_z}{r}+\frac{Gr}{\mathrm{Re}^2}\theta; \quad (2)$$

$$\frac{\partial u_\varphi}{\partial t}+\frac{\partial (u_r u_\varphi)}{\partial r}+\frac{\partial (u_z u_\varphi)}{\partial z}=-\frac{2u_r u_\varphi}{r}+\frac{1}{\mathrm{Re}}\left(\frac{\partial^2 u_\varphi}{\partial r^2}+\frac{1}{r}\frac{\partial u_\varphi}{\partial r}+\frac{\partial^2 u_\varphi}{\partial z^2}-\frac{u_\varphi}{r^2}\right); \quad (3)$$

$$\frac{\partial u_r}{\partial r}+\frac{\partial u_z}{\partial z}+\frac{u_r}{r}=0; \quad (4)$$

$$\frac{\partial\theta}{\partial t}+\frac{\partial(u_r\theta)}{\partial r}+\frac{(\partial u_z\theta)}{\partial z}=\frac{1}{\Pr\cdot\mathrm{Re}}\left(\frac{\partial^2\theta}{\partial r^2}+\frac{1}{r}\frac{\partial\theta}{\partial r}+\frac{\partial^2\theta}{\partial z^2}\right)-\frac{u_r\theta}{r}; \quad (5)$$

$$\frac{\partial C}{\partial t}+\frac{\partial(u_r C)}{\partial r}+\frac{(\partial u_z C)}{\partial z}=\frac{1}{\Pr_D\cdot\mathrm{Re}}\left(\frac{\partial^2 C}{\partial r^2}+\frac{1}{r}\frac{\partial C}{\partial r}+\frac{\partial^2 C}{\partial z^2}\right)-\frac{u_r C}{r}. \quad (6)$$

The dimensionless form of equations is obtained by using the following scales: vortex chamber radius R_0, axial velocity at the input U_0, density at the input ρ_0, temperature and concentration at the input T_0 and C_0 respectively.

The system of equations obtained contains criteria of Reynolds, Prandtl, Schmidt (Prandtl diffusion) and Grashof:

$$\mathrm{Re}=\frac{\rho_0 U_0 R_0}{\mu};\ \Pr=\frac{c_p\mu}{\lambda};\ \Pr_D=\frac{\nu}{D};\ \mathrm{Gr}=\frac{g\beta(T_1-T_0)\rho_0(R_0)^3}{\mu^2}. \quad (7)$$

The dimensionless temperature is determined as $\theta=(T-T_0)/(T_1-T_0)$. In addition, the following notations are introduced: u_r is radial velocity component, u_φ is peripheral velocity component and u_z is axial velocity component, p is pressure.

3.1.2. System of equation in «vortex – stream function» variables. For equations in «vortex – stream function» variables to obtain, one introduce the stream function ψ and vortex Ω [1]:

$$u_r=\frac{1}{r}\frac{\partial\psi}{\partial z},\quad u_z=-\frac{1}{r}\frac{\partial\psi}{\partial r},\quad \Omega=\frac{\partial u_r}{\partial z}-\frac{\partial u_z}{\partial r}. \quad (8)$$

By using these expressions one can transform the set of equations (1) – (6) to the following identical system:

$$\frac{\partial\Omega}{\partial t}+\frac{\partial(u_r\Omega)}{\partial r}+\frac{\partial(u_z\Omega)}{\partial z}=\frac{2u_\varphi}{r}\frac{\partial u_\varphi}{\partial z}+\frac{1}{\mathrm{Re}}\left(\frac{\partial^2\Omega}{\partial r^2}+\frac{1}{r}\frac{\partial\Omega}{\partial r}+\frac{\partial^2\Omega}{\partial z^2}-\frac{\Omega}{r^2}\right)-\frac{Gr}{\mathrm{Re}^2}\frac{\partial\theta}{\partial r}; \quad (9)$$

$$\frac{\partial\psi}{\partial t}=\frac{\partial^2\psi}{\partial r^2}+\frac{\partial^2\psi}{\partial z^2}-\frac{1}{r}\frac{\partial\psi}{\partial r}-r\Omega; \quad (10)$$

$$\frac{\partial u_\varphi}{\partial t}+\frac{\partial(u_r u_\varphi)}{\partial r}+\frac{\partial(u_z u_\varphi)}{\partial z}=-\frac{2u_r u_\varphi}{r}+\frac{1}{\mathrm{Re}}\left(\frac{\partial^2 u_\varphi}{\partial r^2}+\frac{1}{r}\frac{\partial u_\varphi}{\partial r}+\frac{\partial^2 u_\varphi}{\partial z^2}-\frac{u_\varphi}{r^2}\right); \quad (11)$$

$$\frac{\partial\theta}{\partial t}+\frac{\partial(u_r\theta)}{\partial r}+\frac{(\partial u_z\theta)}{\partial z}=\frac{1}{\Pr\cdot\mathrm{Re}}\left(\frac{\partial^2\theta}{\partial r^2}+\frac{1}{r}\frac{\partial\theta}{\partial r}+\frac{\partial^2\theta}{\partial z^2}\right)-\frac{u_r\theta}{r}; \quad (12)$$

$$\frac{\partial C}{\partial t}+\frac{\partial(u_r C)}{\partial r}+\frac{(\partial u_z C)}{\partial z}=\frac{1}{\Pr_D\cdot\mathrm{Re}}\left(\frac{\partial^2 C}{\partial r^2}+\frac{1}{r}\frac{\partial C}{\partial r}+\frac{\partial^2 C}{\partial z^2}\right)-\frac{u_r C}{r}. \quad (13)$$

This system contains five equations instead of six ones since the continuity equation is automatically satisfied by the stream function determination.

3.2. Boundary conditions

For a unique solution to obtain, the boundary conditions in dimensionless form are set. At the input of chamber we have:

$$\frac{\partial u_r}{\partial z}=0, \quad u_\varphi=0, \quad u_z=-1, \quad \theta=0, \quad C=1;$$

at the symmetry axis by $r = 0$ we have

$$u_r=0, \quad u_\varphi=0, \quad \frac{\partial u_z}{\partial r}=0, \quad \frac{\partial \theta}{\partial r}=0, \quad \frac{\partial C}{\partial r}=0;$$

at the walls of chamber there are no-slip and heat insulation conditions (n is normal direction):

$$u_r=0, \quad u_\varphi=0, \quad u_z=0, \quad \frac{\partial \theta}{\partial n}=0, \quad \frac{\partial C}{\partial n}=0,$$

with the exception of rotating surfaces, for which no-slip condition cause an emergence of another criterion, the inverse Rossby number $\mathrm{R\omega} = R_0\omega/U_0$. The value of peripheral velocity at the rotating surface is $u_\varphi = \mathrm{R\omega}\cdot r$. We have $\theta = 1$ and $\partial C/\partial z = kC^m$ at the bottom wall, k is reaction rate constant, m is reaction order. At the chamber output, the Neumann conditions are used for all functions: $\partial/\partial z = 0$.

4. Numerical solution methods

The numerical simulation of a set (1) – (6) was carried out using method of pressure and velocity correction [2]. One can present a system of equations consisting of the momentum equations and the continuity equation in the vector form as follows:

$$\frac{\partial \mathbf{V}}{\partial t}=-\frac{\nabla p}{\rho}+F(\mathbf{V}); \qquad div\mathbf{V}=0. \tag{14}$$

One can obtain the two equations using time-splitting method for the momentum equation:

$$\frac{\mathbf{V}^+-\mathbf{V}^n}{\Delta t}=-\frac{\nabla p^n}{\rho}+F(\mathbf{V}^+,\mathbf{V}^n); \tag{15}$$

$$\frac{\mathbf{V}^{n+1}-\mathbf{V}^+}{\Delta t}=-\frac{\nabla(\delta p)}{\rho}, \tag{16}$$

here δp is the pressure correction which is equal to the difference between the pressures at new n+1st layer of time and at the nth layer known, $\mathbf{V}^+$ is an intermediate grid function. After scalar multiplication of the equation (16) by the gradient and taking into account the solenoidality of velocity on the n+1th layer, we have obtained the Poisson equation for the calculation of pressure correction as follows:

$$\nabla^2(\delta p)=\rho\frac{\nabla\cdot\mathbf{V}^+}{\Delta t}. \tag{17}$$

One can represent this equation as transient Poisson equation for the convenience of calculation:

$$\frac{\partial(\delta p)}{\partial \tau}-\nabla^2(\delta p)=-\rho\frac{\nabla\cdot\mathbf{V}^+}{\partial t}. \tag{18}$$

In equation (18) $\Delta\tau = \mathrm{B}\Delta t$ is the time step performing the role of iteration parameter which value is chosen to ensure the most rapid convergence of problem solution. For the difference scheme to construct, it is used a staggered grid. The set of equations obtained was being solved by using an evolutionary method of relaxation in time. For each equation in the set obtained to solve, an alternating direction implicit scheme was used [3]. This method has the second order of accuracy in time and it is unconditionally stable. Convective and diffusion terms in transport equations for a staggered grid [4] are represented by means the Exponential Scheme [5] based on the control volume method. The alternating direction implicit scheme for a non-staggered grid was used for equations (9) – (13) solving, too. Our own program codes for each method were compiled.

5. Reliability proof

The reliability of results obtained was validated with several ways. Figure 2 shows comparison of graphs of flow characteristics obtained by using different solution methods in sections showed at the right figure. Solid lines are the model predictions in «velocity – pressure» variables, points are the «vortex – stream function» variables method results. A congruence of methods results means the solution right.

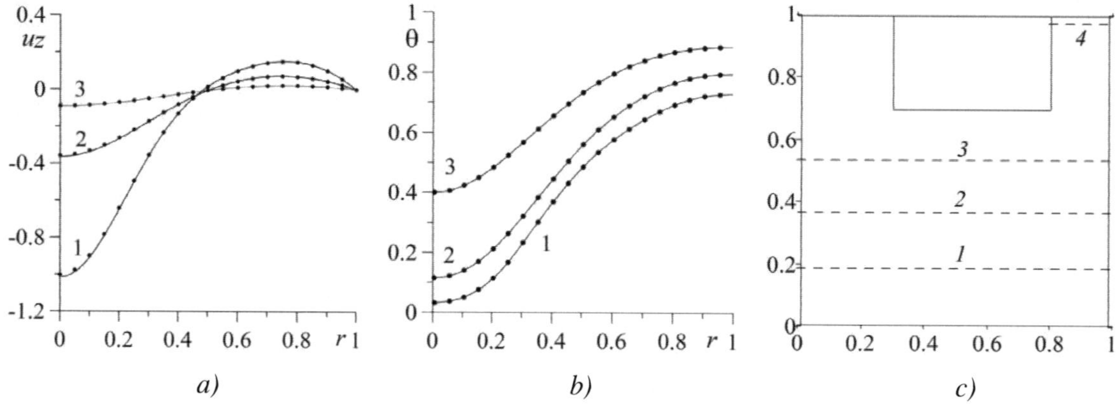

Figure 2. Comparison of solutions obtained by different methods in the sections in a vortex chamber: *a)* axial velocity component; *b)* temperature; *c)* the sections specified

Figure 3*a* illustrates comparison of model prediction for velocity distribution in an annual channel at the vortex chamber output with the exact solution (section 4 at the figure 2*c*) [6].

At the figure 3*b* the comparison of a numerically solving of the classical test problem of a rotating infinite disk in a stationary liquid with the exact data is shown [7]. Figure 3 shows a good agreement of model results with the exact solutions.

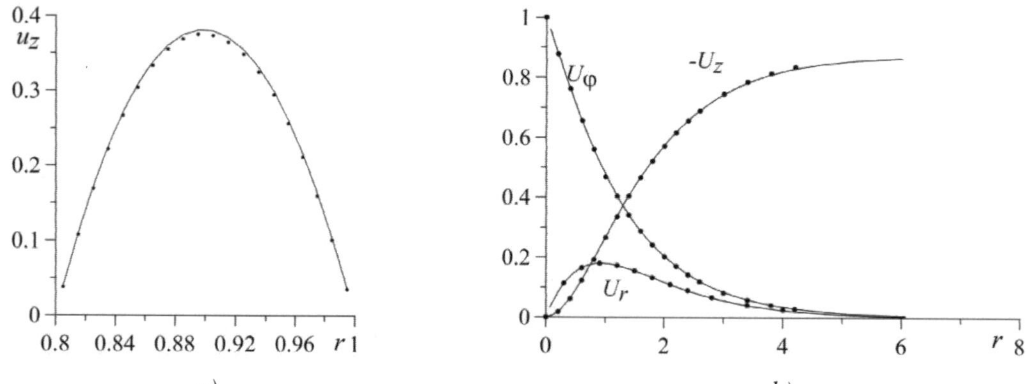

Figure 3. Comparison of the model results with the exact solutions: *a)* for a velocity distribution in an annular channel *b)* for a rotating infinite disk in a stationary liquid

6. Results and Discussion

Numerical calculations were performed for the different reactor operating conditions. The rotating effect upon the hydrodynamics and heat and mass transport in a chemical reactor was studied [8,9]. The results of the numerical solution are presented in figure 4, 5 and 6. Figure 4 demonstrates the stream function for the different rotation degree. One can see that rotation promotes uniform mixing of a matter in the vortex chamber, but too strong rotation can cause precessing core formation.

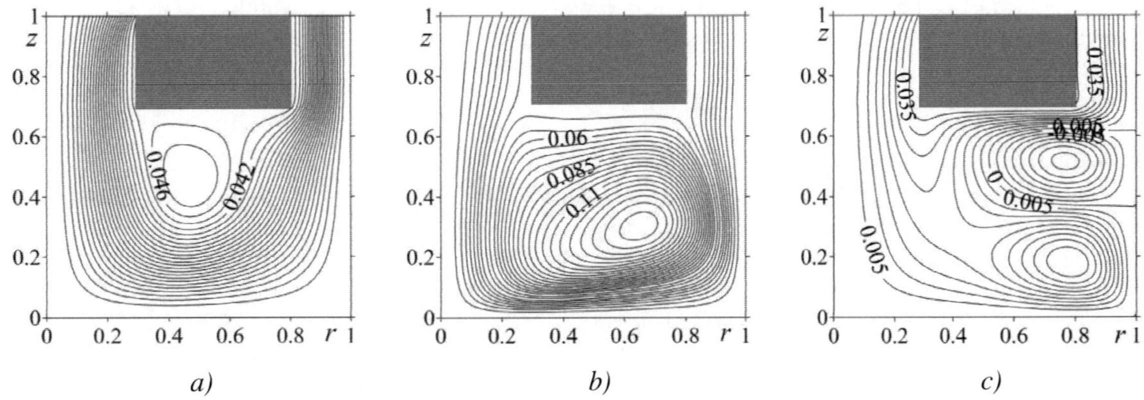

Figure 4. The stream function for the different rotation degree at Re = 50, Gr = 10^5:
*a)*Rω = 0; *b)* Rω = 1; *c)* Rω = 5.

Figure 5 illustrates the solution by rotating effect not taken into account and figure 6 shows the solution with rotation by the same parameters and criteria of the flow considered. The figures presented show that a rotation impacts significantly on the flow hydrodynamics. The distributions of temperature and concentration become more uniform near the susceptor so surface reaction on it proceeds more effective.

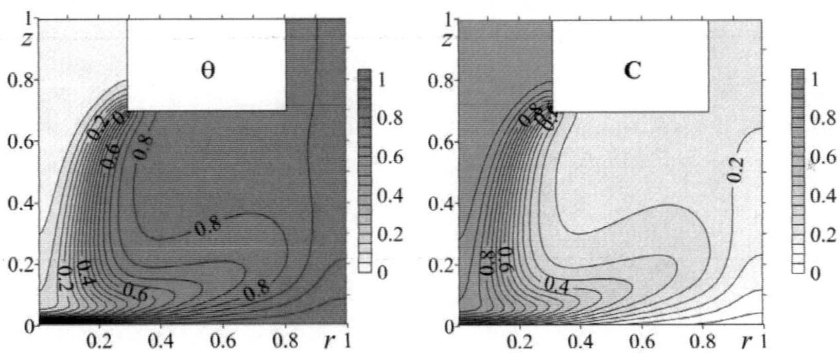

Figure 5. Temperature and concentration distribution in the vortex chamber at Rω = 0. Operating parameters are Re = 50, Pr = 1, Pr_D = 1, Gr = 10^5, k = 5.

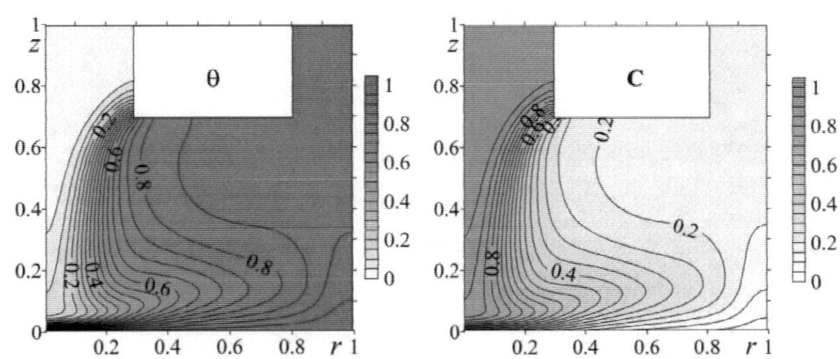

Figure 6. Temperature and concentration distribution in the vortex chamber at Rω = 5. Operating parameters are Re = 50, Pr = 1, Pr_D = 1, Gr = 10^5, k = 5.

7. Conclusions

The fluid motion and heat and mass transfer in a vortex chamber of the chemical reactor has been studied by dint of mathematical simulation. The numerical calculation code compiled predicts the flow character by different operating conditions. The influence of a rotation effect upon the hydrodynamics and heat and mass transport was considered. The rotation results in a more uniform distribution of temperature and matter in the vortex chamber.

References

[1] Shvab A V and Popp M Yu 2014 *Tomsk state university Journal of Mathematics and Mechanics* **28** (Tomsk) pp 90–97
[2] Shvab A V, Gicheva N I and Brendakov V N 2016 Research of interaction of a natural and forced convection in a vortex chamber of the chemical *MATEC Web of Conferences* **92**
[3] Roach P 1972 *Computational Fluid Dynamics* p 376
[4] Chung T J 2002 *Computational Fluid Dynamics* (Cambridge University Press) p 109
[5] Patankar S 1980 *Numerical Heat Transfer and Fluid Flow* (New York: Hemisphere Publishing Corporation) p 86
[6] Petuchov B 1967 *Teploobmen i soprotivlenie pri laminarnom techenii zhidkosti v trubah* (Moskow: Energiya) p 238
[7] Schlichting H and Gestern K 2017 *Boundary-Layer Theory* (Berlin: Springer-Verlag Berlin Heidelberg) chapter 5 p 101–142
[8] Gicheva N I and Dyakov E A 2016 *International Youth Scientific Conference "Current issues of continuum mechanics and celestial mechanics"* (The conference proceedings) ed M Yu Orlov (Tomsk) pp 119–120
[9] Gicheva N I and Dyakov E A 2017 *The Works Tomskogo State Universteta (Series physico-mathematical* **300***)* ed M Yu Orlov (Tomsk) pp 36–41

Acknowledgment

The author appreciates the valuable advices and consultations of his scientific adviser, Doctor of Physics and Mathematics, Professor Alexander Veniaminovich Shvab.

IOP Conf. Series: Journal of Physics: Conf. Series **919** (2017) 012018 doi:10.1088/1742-6596/919/1/012018

Numerical simulation of filtration of mine water from coal slurry particles

E N Dyachenko, N N Dyachenko

Tomsk State University, 36, Lenin Avenue, Tomsk, 634050, Russia

Prof.Dyachenko@gmail.com

Abstract. The discrete element method is applied to model a technology for clarification of industrial waste water containing fine-dispersed solid impurities. The process is analyzed at the level of discrete particles and pores. The effect of filter porosity on the volume fraction of particles has been shown. The degree of clarification of mine water was also calculated depending on the coal slurry particle size, taking into account the adhesion force.

1. Introduction

Clarification of liquid containing fine particles of solid impurities is usually carried out by filtration of the liquid through a layer of porous material, an example of which is a bulk layer. There has been a considerable progress in solving the problems of fluid flow in a medium based on the continuum mechanics assumptions [1; 2]. However, the issues directly related to the formation of the bulk layer structure and the deposition of impurity particles in the channels of the porous medium have been beyond the scope of research. Answering these questions, relying only on the methods of continuum mechanics, seems to be problematic as the process is occurring at the level of discrete particles.

The structure of a bulk layer is modeled as a packing of spherical particles, which is numerically simulated using a discrete element method. The theoretical basis of the discrete element method is presented in [3; 4]. This method allows one to explore media composed of discrete particles. In [5; 6] the discrete element method was applied to modeling the processes of formation of filtration and sedimentation layers of a bulk material. The process of liquid clarification from solid impurities is analogous to the formation of sediments.

This study is devoted to numerical simulation of fluid clarification to remove suspended solid particles using the discrete element method. As a mathematical model, we apply the model of formation of a porous layer of a bulk material [6].

2. Problem formulation

A cube with an edge of unit length is filled with a monodisperse ensemble of sand particles with a radius of R. This results in the formation of a filtration layer with the porosity of $\varepsilon_0 = 0.42$. The volume between the particles, equal to $(1-\varepsilon_0)$, is filled with water. Then impurity particles with a radius of $r = const$ are subsequently let into the cube through its top face with randomly selected coordinates towards the bottom of the container. It is assumed that a particle experiences the force of gravity, $\mathbf{F}_g$, the force of resistance to motion through the medium, $\mathbf{F}_{st}$, and, if contacted with other particles, the adhesion force, F_{ad}. The elastic interaction and friction forces are not taken into account.

The number of particles in the filtration layer is defined as equal to N_R, while that of impurity particles is N_r. An impurity particle with a radius of r_i may interact with both impurity particles and those of the filtration material. The resulting force for the particle r_i can be expressed as

$$\mathbf{F}_i = \mathbf{F}_{g,i} + \mathbf{F}_{st,i} + \sum_{j=1}^{N_r}\left[\mathbf{F}_{ad}(r_i, r_j)\right] + \sum_{j=1}^{N_R}\left[\mathbf{F}_{ad}(r_i, R_j)\right]. \quad (1)$$

For each impurity particle, r_i, the set of motion equations for its center of mass has a form of

$$\begin{cases} \dfrac{d\mathbf{v}_i}{dt} = \dfrac{\mathbf{F}_i}{m_i} \\ \dfrac{d\mathbf{x}_i}{dt} = \mathbf{v}_i, \quad i = 1..N_r \end{cases}, \quad (2)$$

where m_i is the particle mass, **t** is the time, $\mathbf{F}_i$ is the resulting force acting on the particle, $\mathbf{x}_i$ is the spatial coordinate of the particle center, and $\mathbf{v}_i$ is the particle velocity vector.

The force of gravity is written as

$$\mathbf{F}_{gi} = m_i\mathbf{g}, \quad (3)$$

where $\mathbf{g}$ is the acceleration due to gravity.

A significant computing load is related to the calculation of a continuous interparticle medium that limits the number of particles involved in the simulation, so we use a simplified description of such a medium.

The force of resistance to the particle motion through the medium is calculated by the Stokes law:

$$\mathbf{F}_{st,i} = -6\pi\mu r_i \mathbf{v}_i, \quad (4)$$

where μ is the dynamic viscosity of the fluid.

The adhesion force (in the case of electrically neutral fluid between the particles) is determined by the expression [7]:

$$\mathbf{F}_{ad} = \left[\frac{A}{6H_0^2} r^*\right] e = K_{ad} * r^* \vec{e} \quad (5)$$

where A is the Hamaker constant, H_0 is the distance between the surfaces of interacting particles (the intermolecular distance), $r^* = \dfrac{2r_i r_j}{r_i + r_j}$ is the equivalent radius of colliding particles, $\mathbf{e}$ is the unit vector directed from a motionless particle r_j to a depositing one r_i, K_{ad} is the adhesion coefficient.

3. Boundary and initial conditions

Since the problem is solved in terms of discrete elements, the boundaries of the container are represented by the partial surfaces of pseudoparticles with radii of $r = r_b \gg 1$ (where r_b is constant and its value is determined so that within the cube concerned the deviation of the sphere surface from the plane is negligible, for example, $r_b = \infty$) and coordinates of $\mathbf{x}_{bi}$ (i = 1, … 5), namely:

1. (0.5, 0.5, -r_b) for the bottom boundary;
2. (-r_b, 0.5, 0.5) for the left boundary;
3. (r_b+1, 0.5, 0.5) for the right boundary;

4. (0.5, $-r_b+1$, 0.5) for the front boundary;
5. (0.5, r_b+1, 0.5) for the back boundary.

The boundary conditions are formulated as follows:

$$\mathbf{x}_{bi} = const$$
$$|\mathbf{v}_{bi}| = 0, \qquad i = 1, \ldots 5. \tag{6}$$

The initial conditions for the set of particles N_r are the following:

$$\mathbf{x}_{i,initial} = random; \tag{7}$$
$$\mathbf{v}_{i,initial} = const,$$

where *random* is an arbitrary real number, from 0 to 1, uniformly distributed. The initial coordinates of the particles, although being random, satisfy the no-intersection condition for any *i* and *j*, namely:

$$|\mathbf{x}_i - \mathbf{x}_j| > r_i + r_j.$$

The most complete mathematical formulation of the problem is presented in [6].

4. Numerical simulation

To solve the problem numerically, we integrate the set of equations (2) taking into account expressions (1, 3-5), boundary (6) and initial (7) conditions. When considering the problem of clarifying industrial waste waters from solid particles by means of bulk filters, the following assumption is used. When an impurity particle collides with a previously deposited one or a particle of the filtration material, it will either adhere to one of them or change the vector of its velocity and continue moving along the surface of contact.

A diagram of the forces acting on a particle r_i while colliding with a particle r_j (note that r_j can be both a particle of the filtering material and a previously deposited impurity particle) is shown in Fig. 1.

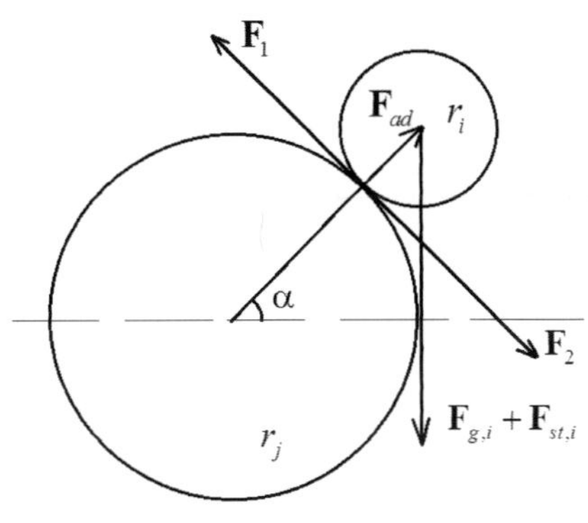

Figure 1. Diagram of interaction between particles

The condition for a particle r_i to stop its motion, i.e. to adhere to one of the particles mentioned above, is determined by the balance between forces holding it on the surface of a particle r_j and forces tending to continue its motion.

The holding force is $|\mathbf{F}_1| = |\mathbf{F}_{ad}|$, whereas the moving force is $|\mathbf{F}_2| = |\mathbf{F}_{g,i} + \mathbf{F}_{st,i}|$.

5. Results and discussion

The input parameters were the following: spherical sand particles with a radius of R = $5 \cdot 10^{-4} m$ (in the dimensionless form, R = 0.005, which was derived by dividing the initial radius by the cube width equal to 1); the packing porosity of the sand layer, $\varepsilon_0 = 0.42$; spherical coal particles with a radius of *r* ranged from *r* = *R/10* to *r* = *R/35*; the adhesion coefficient, $K_{ad} = 1{\cdot}10^{2}\, H/m$; the distance between the surfaces of interacting particles, $H_0 = 4{\cdot}10^{-8}\ m$; the viscosity of water, $\mu = 10^{-3}\, Pa \cdot s$; the density of coal, $\rho_y = 1.7 \cdot 10^3\, kg/m^3$.

Fig. 2 shows the results of calculating the dependence of the porosity in the bidispersed packing (R/r = 30) on the volume fraction of fine particles (V_r / V), where V_r is the dimensionless volume of fine particles and *V* is the volume of the unit cube. Curve 1 corresponds to the calculation in the absence of the adhesion force (K_{ad} = 0), Curve 2 corresponds to the calculation taking into account the adhesion force between impurity particles and sand particles as well as between impurity particles themselves.

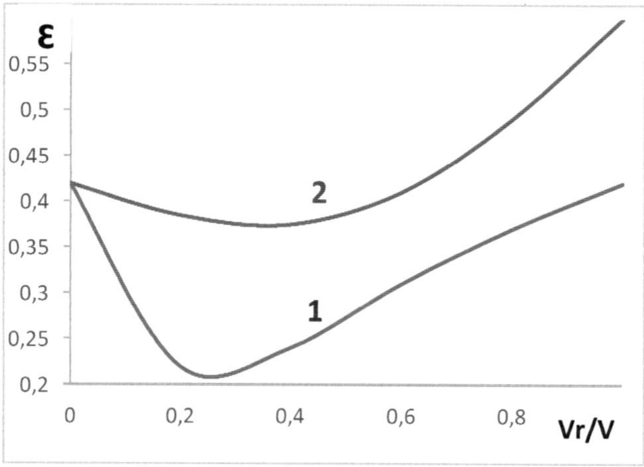

Figure 2. Dependence of porosity on volume fraction of impurity particles

In the absence of adhesive interaction at points $V_r / V = 0$ or 1 the porosity of the packing is equal to 0.42, as at these points there is a set of monodisperse particles (either pure sand or impurity particles).

When adhesive interaction is taken into account, at the point $V_r / V = 0$ the porosity equals $\varepsilon = 0.42$, whereas at the point $V_r / V = 1$ it is $\varepsilon = 0.58$ and determined by the adhesion force between impurity particles.

Fig. 3 depicts the results of calculating the dependence of the fluid clarification degree, M, on the particle size ratio, R/r. Here, M $= (n_1 / n_2) \cdot 100\ \%$, n_1 is the number of impurity particles retained by the filter, n_2 is the number of impurity particles passed from the upper face of the cube. Curve 1 corresponds to the calculation without adhesion ($K_{ad} = 0$) and Curve 2 shows the calculation taking adhesive interaction into account. As follows, in the absence of adhesion the impurity particles whose size is *r* = *R*/20 are almost not trapped by the sand layer with a thickness of 0.1 m. Taking into account adhesive interaction results in trapping 58% of the impurity particles whose size is *r* = *R*/10 and 2% for the particles whose size is r = R/35 using the sand layer with a thickness of 0.1 m.

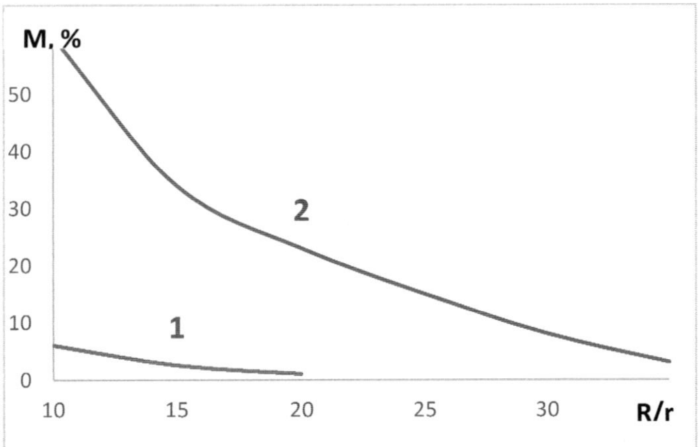

Figure 3. Dependence of clarification degree on sand-to-impurity particle size ratio

To improve the degree of clarification of water containing particles of a smaller size, it is necessary to increase the thickness of the sand layer or to use appropriate surfactants to enhance the adhesion interaction between particles.

As seen in Fig. 4, the solid line shows the results of calculating the porosity of the layer along the container when the adhesive force between particles, a particle and the wall of the container was not taken into account. At the wall of the container the porosity is equal to unity because a particle contacts the surface at this point. Further, the porosity is observed to be fluctuating up to five particle radii. The dotted line in the figure shows the experimental values of the porosity for the bulk layer borrowed from [8]. Comparison of the calculated and experimental results shows their qualitative agreement. When filtering mine waters, the space between sand particles is filled with smaller coal slurry particles, with the thickness of the near-wall layer being determined by the latter ones. In using sand filters in the coal industry, the influence of the near-wall layer can be neglected since its thickness is much smaller than that of the filter.

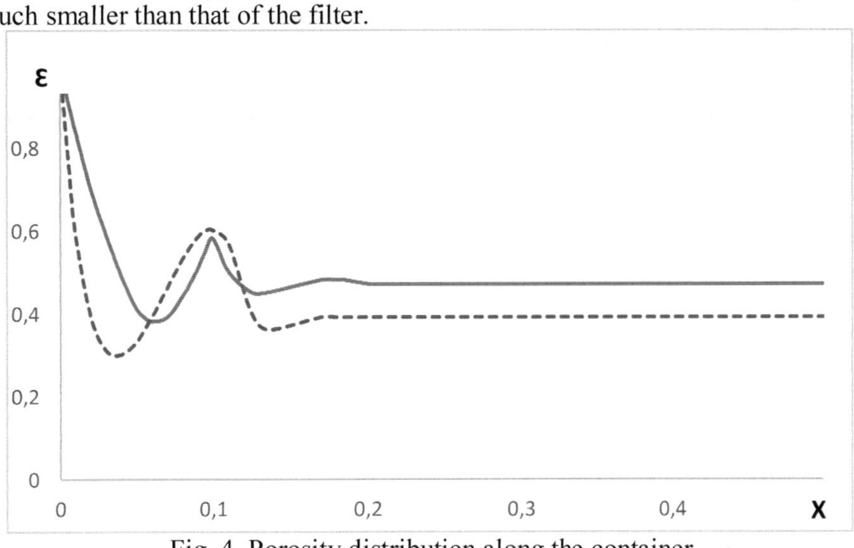

Fig. 4. Porosity distribution along the container

The filter is governed by the total pressure drop in the deposited, blocking and pure sand layers [9]. The layer of sand lying below the blocking one performs no filtering function, but it increases the pressure drop, thereby increasing the energy consumption of sewage treatment plants. Therefore, solving the task of clarifying fluid from a polydisperse ensemble of solid particles, it is necessary to consider this factor.

Numerical implementation of the problem was performed using the computer program PORA [10] on the multiprocessor cluster SKIF-CYBERIA.

6. Conclusions
The calculation method proposed allows one to consider the process of filtration of fluid at the level of discrete particles and pores, to determine the degree of clarification of mine waste water containing fine coal slurry particles and to optimize the degree of filter loading depending on the filtrate parameters and the operating mode of treatment facilities. The results obtained can be used for designing bulk filters.

References
[1] Zhuzhikov V.A. Fil'trovanie: teoriya i praktika razdeleniya suspensiy [Filtration: theory and practice of suspension separation]. Moscow: Khimiya, 1971. 441 p.
[2] Wakeman R.J., Tarleton E.S. Filtration: equipment selection, modeling and process simulation. Oxford, UK; New York: Elsevier Advanced Technology, 1999. 446 p.
[3] Zhu H.P., Zhou Z.Y., Yang R.Y., Yu A.B. Discrete particle simulation of particulate systems: theoretical development // Chemical Engineering Science. 2007. Vol. 62, is. 13. P. 3378-3396.
[4] Zhu H.P., Zhou Z.Y., Yang R.Y., Yu A.B. Discrete particle simulation of particulate systems: a review of major applications and findings // Chemical Engineering Science. 2008. Vol. 63, is. 23. P. 5728-5770.
[5] Nesse Th., Dueck J., Dyachenko E. Simulation of filter cake porosity in solid/liquid separation // Powder Technology. 2009. Vol. 193. P. 332-335.
[6] Dyachenko E.N., Dueck J. G. Computer simulation of porous layers based on the method of discrete elements // Journal of Engineering Physics and Thermophysics. 2013. Vol. 86, is. 6. P. 1315-1327.
[7] Israelachvili J. Intermolecular and surface forces. San Diega: Academic Press, 1955. 450 p.
[8] Brayer H. Grundlagen der Einplazen and Mehrphasenstromunge. Aarau, Switzerland: Sauerlander, 1971. 955 p.
[9] Vasenin I.M., Dyachenko N.N., Dyachenko L.I. Osvetlenie shakhtnykh vod na sloe peska [Clarification of mine water by filtering it through a sand bed] // Isvestiya Vysshikh Uchebnykh Zavedenii. Gornyi Zhurnal. 2004. Vol. 6. P. 50-54.
[10] Dyachenko E.N., Minkov L.L. Raschet formirovaniya nasypnogo sloya iz polidispersnykh chastits staticheskim metodom [Calculation of bulk layer formation from polydisperse particles by the static method]. Computer Program Patent, no. 2013615471 issued by Pospatent (June 10, 2013).

IOP Conf. Series: Journal of Physics: Conf. Series **919** (2017) 012019 doi:10.1088/1742-6596/919/1/012019

On improvement of the series convergence in the problem of the vibrations of orhotropic rectangular prism

A D Lyashko

V I Vernadsky Crimean Federal University, 4 Vernadskogo Prospekt, Simferopol, Crimea, Russia

E-mail: knightla@yandex.ru

Abstract. A new analytical presentation of the solution for steady-state oscillations of orthotropic rectangular prism is found. The corresponding infinite system of linear algebraic equations has been deduced by the superposition method. A countable set of precise eigenfrequencies and elementary eigenforms is found. The identities are found which make it possible to improve the convergence of all the infinite series in the solution of the problem. All the infinite series in presentation of solution are analytically summed up. Numerical calculations of stresses in the rectangular orthotropic prism with a uniform along the border and harmonic in time load on two opposite faces have been performed.

1. Introduction

Orthotropic prisms and plates are among the most widespread and modern elements of structures in engineering and construction. Many researches [1-6] applied different variants of the superposition method to problems of vibrations of rectangular prisms and plates. Dynamic behavior of thin orthotropic plates is studied using the Rayleigh-Ritz method most fully in [7]. One of the common numerical methods applied to the study of these problems is finite element method [8].

Grinchenko and Meleshko considered the problem of vibration of isotropic rectangular prism in [1]. They constructed the displacements as sum of two solutions each of which allows satisfying arbitrary conditions on displacement and stress on corresponding pair of parallel sides of the prism. The generalization of this approach for orthotropic plates is considered by Papkov [2]. Even in case of isotropic prisms the use of obtained solutions is difficult because of logarithmic singularities in normal stresses near the corners. The convergence of series in expressions for stresses deteriorates near the sides of the prism. Because of this convergence of series is being improved in one way or another. Additionally a countable set of elementary eigenforms for a countable set of ratios of prism's sides is specified in [1] called Lamé modes.

In this paper all the series in presentation of solution are analytically summed up. The convergence of series is improved with the help of theory of regular infinite system of linear algebraic equations. The generalization of Lamé modes for orthotropic prisms is obtained.

2. Formulation of the problem and analytical presentation of solution

The plane deformation of orthotropic prism, cross-section of which is a rectangle with sides $2a \times 2b$, is considered. Prism's sides are parallel to X and Y axes. The origin of coordinate system is at the prism's mid-point. The basis vectors are normal to planes of symmetry of orthotropic material. The equations of motion for an elastic solid are written as follows:

$$\frac{\partial\tilde{\sigma}_{xx}}{\partial x}+\frac{\partial\tilde{\tau}_{xy}}{\partial y}=\rho\frac{\partial^2\tilde{u}}{\partial t^2},\quad \frac{\partial\tilde{\tau}_{xy}}{\partial x}+\frac{\partial\tilde{\sigma}_{yy}}{\partial y}=\rho\frac{\partial^2\tilde{v}}{\partial t^2}. \tag{1}$$

The stress-displacement relations are given by the generalized Hooke's law:

$$\frac{1-\nu_1\nu_2}{E_1}\tilde{\sigma}_{xx}=\frac{\partial\tilde{u}}{\partial x}+\nu_2\frac{\partial\tilde{v}}{\partial y},\quad \frac{1-\nu_1\nu_2}{E_2}\tilde{\sigma}_{yy}=\nu_1\frac{\partial\tilde{u}}{\partial x}+\frac{\partial\tilde{v}}{\partial y},\quad \frac{\tilde{\tau}_{xy}}{G}=\frac{\partial\tilde{u}}{\partial y}+\frac{\partial\tilde{v}}{\partial x}, \tag{2}$$

where ν_1, ν_2, E_1, E_2 and G are respectively Poisson's ratios, Young's moduli in direction of X and Y axes and shear modulus.

Steady-state vibration of rectangular prism, symmetric about both axes, under load normal to its sides is considered.

$$\left.\frac{1-\nu_1\nu_2}{E_1}\tilde{\sigma}_{xx}\right|_{x=\pm a}=f(y)\sin(\omega t),\quad \left.\frac{\tilde{\tau}_{xy}}{G}\right|_{x=\pm a}=0, \tag{3}$$

$$\left.\frac{1-\nu_1\nu_2}{E_2}\tilde{\sigma}_{yy}\right|_{y=\pm b}=g(x)\sin(\omega t),\quad \left.\frac{\tilde{\tau}_{xy}}{G}\right|_{x=\pm b}=0. \tag{4}$$

Amplitudes of displacements and stresses are introduced as follows:

$$\tilde{u}=u\sin(\omega t),\quad \tilde{v}=v\sin(\omega t), \tag{5}$$

$$\tilde{\sigma}_{xx}=\sigma_{xx}\sin(\omega t),\quad \tilde{\sigma}_{yy}=\sigma_{yy}\sin(\omega t),\quad \tilde{\sigma}_{xy}=\sigma_{xy}\sin(\omega t). \tag{6}$$

Analytical presentation of the solution is built based on the superposition method outlined in [1]. Functions in the solutions and part of coefficients are chosen to satisfy differential equations and boundary conditions for shear stress. To satisfy the boundary conditions for normal stress an infinite system of linear equations is constructed. Under this approach a new analytical presentation of solution for amplitudes of displacements was obtained in the form of the following trigonometric Fourier series:

$$\begin{aligned} u=&\frac{N_1}{b}\frac{Y_0}{\Omega_{12}^2}\frac{\sin\Omega_{11}x}{\sin\Omega_{11}a}+\frac{2N_2}{a}\sum_{n=1}^{\infty}X_n(-1)^n\left(B_{11}(n,\alpha,p)\frac{\cosh p_{1n}y}{\sinh p_{1n}b}+B_{12}(n,\alpha,p)\frac{\cosh p_{2n}y}{\sinh p_{2n}b}\right)\sin\alpha_n x-\\ &\frac{2N_1}{b}\sum_{m=1}^{\infty}Y_m(-1)^m\left(C(\beta_m,q_{1m},E_2)\,B_{21}(m,\beta,q)\frac{\sinh q_{1m}x}{\sinh q_{1m}a}+\right.\\ &\left.C(\beta_m,q_{2m},E_2)\,B_{22}(m,\beta,q)\frac{\sinh q_{2m}x}{\sinh q_{2m}a}\right)\cos\beta_m y, \end{aligned} \tag{7}$$

$$\begin{aligned} v=&-\frac{N_2}{a}\frac{X_0}{\Omega_{12}}\frac{\sin\Omega_{12}y}{\sin\Omega_{12}b}-\frac{2N_1}{b}\sum_{m=1}^{\infty}Y_m(-1)^m\left(B_{21}(m,\beta,q)\frac{\cosh q_{1m}x}{\sinh q_{1m}a}+B_{22}(m,\beta,q)\frac{\cosh q_{2m}x}{\sinh q_{2m}a}\right)\sin\beta_m y+\\ &\frac{2N_2}{a}\sum_{n=1}^{\infty}X_n(-1)^n\left(C(\alpha_n,p_{1n},E_1)\,B_{11}(n,\alpha,p)\frac{\sinh p_{1n}y}{\sinh p_{1n}b}+\right.\\ &\left.C(\alpha_n,p_{2n},E_1)\,B_{12}(n,\alpha,p)\frac{\sinh p_{2n}y}{\sinh p_{2n}b}\right)\cos\alpha_n x. \end{aligned} \tag{8}$$

Here

$$\Omega_{11}=\Omega_2\sqrt{G(1-\nu_1\nu_2)/E_1}\,,\ \Omega_{12}=\Omega_2\sqrt{G(1-\nu_1\nu_2)/E_2}\,,\ \Omega_2=\pi\Omega/2,\ \Omega=\frac{2\omega}{\pi}\sqrt{\frac{\rho}{G}}\,,$$

$$B_{jk}(n,\gamma,r)=\frac{r_{k,n}\gamma_n(r_{k,n}^2\nu_{\hat{j}}+\gamma_n^2-\Omega_{1j}^2)}{(r_{\hat{k},n}^2-r_{k,n}^2)(\gamma_n^2-\Omega_{1j}^2)(\Omega_{1j}^2-\gamma_n^2(1-\nu_1\nu_2)/(1+\nu_j))}\,,\ \hat{j}=3-j\,,$$

$$N_j=\sqrt{\frac{1+\nu_j}{1+\nu_{\hat{j}}}}\,,C(\gamma,r,E)=\frac{(1-\nu_1\nu_2)(r^2+\Omega_2^2)-\gamma^2E/G}{(1+\nu_1E_2/G-\nu_1\nu_2)r\,\gamma}\,,\ \alpha_n=\frac{n\pi}{a}\,,\ \beta_m=\frac{m\pi}{b}\,.$$

p_{1n}, p_{2n}, q_{1m}, q_{2m} are the roots of the following biquadratic equations:

$$p^4+\left(\Omega_{12}^2+\Omega_2^2-\left(\frac{E_1}{G}-2\nu_1\right)\alpha_n^2\right)p^2+\frac{\nu_1}{\nu_2}(\alpha_n^2-\Omega_{11}^2)(\alpha_n^2-\Omega_2^2)=0\,, \tag{9}$$

$$q^4+\left(\Omega_{11}^2+\Omega_2^2-\left(\frac{E_2}{G}-2\nu_2\right)\beta_m^2\right)q^2+\frac{\nu_2}{\nu_1}(\beta_m^2-\Omega_{12}^2)(\beta_m^2-\Omega_2^2)=0\,. \tag{10}$$

Shear stress vanishes on the faces of the prism due to the choice of the presentation of the solution. Substitution of expressions for normal stresses into the boundary conditions after a known procedure of the superposition method leads to an infinite system of linear algebraic equations with respect to indeterminate coefficients X_k, Y_k $(k=0,1,2,\ldots)$. Constant terms in the infinite system are proportional to coefficients of Fourier series expansion of boundary conditions for normal stresses:

$$f(y)=f_0+\sum_{m=1}^{\infty}(-1)^m f_m\cos\beta_m y\,,\ g(x)=g_0+\sum_{n=1}^{\infty}(-1)^n g_n\cos\alpha_n x\,. \tag{11}$$

The infinite system for the solution presentation (7), (8) is as follows:

$$\begin{aligned}
&X_0\cot(\Omega_{12}b)N_2^2=\frac{\nu_2}{\Omega_{12}b}Y_0+\frac{2}{b}\sum_{m=1}^{\infty}Y_m\frac{\Omega_{12}^3\nu_2}{\left(\beta_m^2\frac{1-\nu_1\nu_2}{1+\nu_2}-\Omega_{12}^2\right)\left(\beta_m^2-\Omega_{12}^2\right)}-a\Omega_{11}\sqrt{\frac{\nu_2}{\nu_1}}N_2g_0\,,\\
&Y_0\cot(\Omega_{11}a)N_1^2=\frac{\nu_1}{\Omega_{11}a}X_0+\frac{2}{a}\sum_{n=1}^{\infty}X_n\frac{\Omega_{11}^3\nu_1}{\left(\alpha_n^2\frac{1-\nu_1\nu_2}{1+\nu_1}-\Omega_{11}^2\right)\left(\alpha_n^2-\Omega_{11}^2\right)}+b\Omega_{11}\sqrt{\frac{\nu_1}{\nu_2}}N_1f_0\,,\\
&X_n\delta_nN_2^2=\frac{1}{b}\frac{\nu_2}{\alpha_n^2-\Omega_{11}^2}Y_0+\frac{2}{b}\sum_{m=1}^{\infty}Y_m\frac{A_{n,m}}{\left(\alpha_n^2+q_{1m}^2\right)\left(\alpha_n^2+q_{2m}^2\right)\left(\beta_m^2\frac{1-\nu_1\nu_2}{1+\nu_2}-\Omega_{12}^2\right)}+\frac{a}{2}N_2g_n\quad(n=1,2,\ldots)\,,\\
&Y_m\delta_m^*N_1^2=\frac{1}{a}\frac{\nu_1}{\beta_m^2-\Omega_{12}^2}X_0+\frac{2}{a}\sum_{n=1}^{\infty}X_n\frac{A_{n,m}}{\left(\beta_m^2+p_{1n}^2\right)\left(\beta_m^2+p_{2n}^2\right)\left(\alpha_n^2\frac{1-\nu_1\nu_2}{1+\nu_1}-\Omega_{11}^2\right)}+\frac{b}{2}N_1f_m\quad(m=1,2,\ldots)\,,
\end{aligned} \tag{12}$$

where

$$A_{n,m}=\beta_m^2(\alpha_n^2(1-\nu_1\nu_2)-\nu_1\Omega_{11}^2)-\nu_1\Omega_{11}^2(\alpha_n^2-\Omega_2^2)\,,$$

$$\delta_n=B_{11}(n,\alpha,p)(\nu_1\alpha_n+C(\alpha_n,p_{1n},E_1)p_{1n})\cot p_{1n}b+B_{12}(n,\alpha,p)(\nu_1\alpha_n+C(\alpha_n,p_{2n},E_1)p_{2n})\cot p_{2n}b\,,$$

$$\delta_m^*=B_{21}(m,\beta,q)(\nu_2\beta_m+C(\beta_m,q_{1m},E_2)q_{1m})\cot q_{1m}a+B_{22}(m,\beta,q)(\nu_2\beta_m+C(\beta_m,q_{2m},E_2)q_{2m})\cot q_{2m}a\,.$$

3. Lamé modes and infinite system study

The presentation of the solution (7), (8) differs from other known presentations [2] by the fact that a unit solution ($X_k = Y_k = 1$) of infinite system (12) corresponds to the following elementary analytical solution of the steady-state vibration of orthotropic prism:

$$u = \sqrt{\frac{1+\nu_1}{1-\nu_1\nu_2}}\frac{1}{\Omega_{12}}\frac{\cos(\Omega_2 y\sqrt{G(1+\nu_2)/E_2})\sin(\Omega_2 x\sqrt{G(1+\nu_1)/E_1})}{\sin(\Omega_2 b\sqrt{G(1+\nu_2)/E_2})\sin(\Omega_2 a\sqrt{G(1+\nu_1)/E_1})}, \tag{13}$$

$$v = -\sqrt{\frac{1+\nu_2}{1-\nu_1\nu_2}}\frac{1}{\Omega_{11}}\frac{\sin(\Omega_2 y\sqrt{G(1+\nu_2)/E_2})\cos(\Omega_2 x\sqrt{G(1+\nu_1)/E_1})}{\sin(\Omega_2 b\sqrt{G(1+\nu_2)/E_2})\sin(\Omega_2 a\sqrt{G(1+\nu_1)/E_1})}, \tag{14}$$

$$\frac{1-\nu_1\nu_2}{E_1}\sigma_{xx} = \sqrt{\frac{\nu_2}{\nu_1}}\frac{\cos(\Omega_2 y\sqrt{G(1+\nu_2)/E_2})\cos(\Omega_2 x\sqrt{G(1+\nu_1)/E_1})}{\sin(\Omega_2 b\sqrt{G(1+\nu_2)/E_2})\sin(\Omega_2 a\sqrt{G(1+\nu_1)/E_1})}, \tag{15}$$

$$\frac{1-\nu_1\nu_2}{E_1}\sigma_{yy} = -\frac{\nu_1}{\nu_2}\frac{1-\nu_1\nu_2}{E_2}\sigma_{xx}, \quad \frac{\tau_{xy}}{G} = 0. \tag{16}$$

For the following discrete set of frequencies and prism sides ratios:

$$\Omega a = (1+2k)\sqrt{\frac{E_1}{G(1+\nu_1)}}, \quad \frac{b}{a} = \frac{1+2l}{1+2k}\sqrt{\frac{1+1/\nu_1}{1+1/\nu_2}}, \quad k,l \in Z_+ \tag{17}$$

stresses (15), (16) vanish at prism border. Therefore, analytical presentation of solution (7), (8) proposed here leads not only to the exact elementary particular solution of infinite system of linear algebraic equations and the exact elementary solution of the steady-state vibration of orthotropic prism, but also to the infinite discrete set of eigenfrequencies and elementary eigenforms of rectangular orthotropic prism. In case of isotropic prism these eigenfrequencies and elementary eigenforms were found in another way in monography [1] and called Lamé modes.

Knowing the unit solution of infinite system (12) significantly simplifies study of its regularity. Obtaining all Fourier series coefficients f_k, g_k for stresses (15), (16) and substituting them alongside with the unit solution ($X_k = Y_k = 1$) into infinite system (12) produces identities from which follow simple exact formulas for sum of coefficients in each equation of infinite system:

$$S_{x0} = 1 - \frac{\tan\Omega_{12} b}{\tan(\Omega_2 b\sqrt{\frac{G(1+\nu_2)}{E_2}})}\sqrt{\frac{1-\nu_1\nu_2}{1+\nu_2}}, \quad S_{y0} = 1 - \frac{\tan\Omega_{11} a}{\tan(\Omega_2 a\sqrt{\frac{G(1+\nu_1)}{E_1}})}\sqrt{\frac{1-\nu_1\nu_2}{1+\nu_1}} \tag{18}$$

$$S_{xn} = 1 - \frac{1}{\delta_n}\cot\left(\Omega_2 b\sqrt{\frac{G(1+\nu_2)}{E_2}}\right)\sqrt{\frac{1-\nu_1\nu_2}{1+\nu_2}}\frac{\Omega_{12}}{\alpha_n^2\frac{1-\nu_1\nu_2}{1+\nu_1} - \Omega_{11}^2}, \tag{19}$$

$$S_{ym} = 1 - \frac{1}{\delta_m^*}\cot\left(\Omega_2 a\sqrt{\frac{G(1+\nu_1)}{E_1}}\right)\sqrt{\frac{1-\nu_1\nu_2}{1+\nu_1}}\frac{\Omega_{11}}{\beta_m^2\frac{1-\nu_1\nu_2}{1+\nu_2} - \Omega_{12}^2}. \tag{20}$$

For sufficiently low vibration frequency all coefficients are positive and these expressions are less than 1. Thus infinite system (12) is regular in this case. Regularity of this infinite system is further studied in article [9].

Finally, the presence of a solution of the infinite system that tends to a nonzero constant allows us to make an assumption that in general case solutions of infinite system follow the asymptotic law:

$$\lim_{k\to\infty} X_k = \lim_{k\to\infty} Y_k = L. \tag{21}$$

In view of this assumption it is possible to apply method [1] of «improved reduction». In accordance with this method an assumption is made that starting with some number P all variables are approximately equal to the limit L. The following change of variables:

$$X_k = \tilde{X}_k + \breve{X}_k L,\; Y_k = \tilde{Y}_k + \breve{Y}_k L \;\; (k = 0,1,\dots P-1), \tag{22}$$

$$X_k = L,\; Y_k = L \;\; (k = P, P+1,\dots) \tag{23}$$

results in two infinite systems of linear equations with the same matrix with respect to $\tilde{X}_k, \tilde{Y}_k$ and $\breve{X}_k, \breve{Y}_k$. To find the limit L an approach analogous to the one described in [1] for isotropic case is employed. The relation between the variables of solutions included in the infinite system is obtained by equating the difference of the normal stresses at the corner point to the value known from boundary conditions (3), (4):

$$\frac{1-\nu_1\nu_2}{E_1}\sigma_{xx}(a,b) - \frac{1-\nu_1\nu_2}{E_2}\sigma_{yy}(a,b) = f(b) - g(a). \tag{24}$$

After substitution of the expressions for normal stresses into equation (24) the expression linear with respect to L is obtained. The series in this expression converge quickly therefore they are mainly determined by the values of the first variables.

4. Improvement of the series convergence and discussion of results

Direct evaluations show that series in normal stresses diverge in corner points. The asymptotical law (21) and known unit solution allows improving series convergence at the border and within the prism. In order to do so Krylov's method of convergence improvement [10] is used. The limit L is added to and subtracted from each indeterminate coefficient. The added terms are separated and summed up with the help of identity for the unit solution. Having done this procedure for normal stress σ_{xx} the following expression is obtained:

$$\begin{aligned}
\frac{1-\nu_1\nu_2}{E_1}\sigma_{xx} = {}& \frac{N_1}{b}\frac{(Y_0-L)\Omega_{11}}{\Omega_{12}^2}\frac{\cos\Omega_{11}x}{\sin\Omega_{11}a} - \frac{N_2}{a}\frac{(X_0-L)\Omega_{12}}{\Omega_{11}^2}\frac{\cos\Omega_{12}y}{\sin\Omega_{12}b} + \\
& \frac{2N_2}{a}\sum_{n=1}^{\infty}(X_n-L)(-1)^n\Big(B_{11}(n,\alpha,p)(\alpha_n + C(\alpha_n,p_{1n},E_1)p_{1n}\nu_2)\frac{\cosh p_{1n}y}{\sinh p_{1n}b} + \\
& B_{12}(n,\alpha,p)(\alpha_n + C(\alpha_n,p_{2n},E_1)p_{2n}\nu_2)\frac{\cosh p_{2n}y}{\sinh p_{2n}b}\Big)\cos\alpha_n x - \\
& \frac{2N_1}{a}\sum_{m=1}^{\infty}(Y_m-L)(-1)^m\Big(B_{21}(m,\beta,q)(C(\beta_m,q_{1m},E_2)q_{1m} + \beta_m\nu_2)\frac{\cosh q_{1m}x}{\sinh q_{1m}a} + \\
& B_{22}(m,\beta,q)(C(\beta_m,q_{2m},E_2)q_{2m} + \beta_m\nu_2)\frac{\cosh q_{2m}x}{\sinh q_{2m}a}\Big)\cos\beta_m y + \\
& L\sqrt{\frac{\nu_2}{\nu_1}}\frac{\cos(\Omega_2 y\sqrt{G(1+\nu_2)/E_2})\cos(\Omega_2 x\sqrt{G(1+\nu_1)/E_1})}{\sin(\Omega_2 b\sqrt{G(1+\nu_2)/E_2})\sin(\Omega_2 a\sqrt{G(1+\nu_1)/E_1})}.
\end{aligned} \tag{25}$$

This procedure was employed for a rectangular orthotropic prism with a=3, b=2, $\nu_1 = 0{,}072$; $\nu_2 = 0{,}036$; $E_1 = 1{,}2$; $E_2 = 0{,}6$; $G = 0{,}07$ and the frequency of forced vibrations $\Omega = 1{,}47$ for the following boundary condition:

$$\frac{1-\nu_1\nu_2}{E_1}\sigma_{xx}\Big|_{x=\pm a} = 1,\; \frac{1-\nu_1\nu_2}{E_2}\sigma_{yy}\Big|_{y=\pm b} = \frac{\tau_{xy}}{G}\Big|_{x=\pm a} = \frac{\tau_{xy}}{G}\Big|_{y=\pm b} = 0. \tag{26}$$

The corresponding infinite system of equations is solved and the stresses along prism's sides are computed. The results of computation with P=21 are shown in table 1 and table 2 for stresses σ_{xx} and σ_{yy} accordingly.

Table 1. Stress σ_{xx} along the side x=a with P=21.

y/b	0,0	0,2	0,4	0,6	0,8	1,0
Without convergence improvement	0,9389	0,9391	0,9399	0,9426	0,9571	4,9553
With convergence improvement	1,0003	1,0003	1,0003	1,0003	1,0004	0,9646

Table 2. Stress σ_{yy} along the side y=b with P=21.

x/a	0,0	0,2	0,4	0,6	0,8	1,0
Without convergence improvement	0,0942	0,0940	0,0928	0,0891	0,6899	–7,2397
With convergence improvement	0,0000	0,0000	0,0000	0,0001	0,0003	–0.0354

More accurate the boundary condition satisfaction may be achieved by increasing the number of equations in systems for $\tilde{X}_k, \tilde{Y}_k$ and $\breve{X}_k, \breve{Y}_k$, i.e. increasing the number of indeterminate coefficients that are not considered approximately equal to L. The corresponding stresses computed with P=101 are presented in table 3 and table 4. Without convergence improvement the boundary conditions near the corners are not satisfied independently of P due to logarithmic singularity in the series.

Table 3. Stress σ_{xx} along the side x=a with P=101.

y/b	0,0	0,2	0,4	0,6	0,8	1,0
Without convergence improvement	0,9875	0,9875	0,9875	0,9876	0,9882	4,9787
With convergence improvement	1,0000	1,0000	1,0000	1,0000	1,0000	1,0004

Table 4. Stress σ_{yy} along the side y=b with P=101.

x/a	0,0	0,2	0,4	0,6	0,8	1,0
Without convergence improvement	0,0193	0,0193	0,0193	0,0191	0,0182	–7,1868
With convergence improvement	0,0000	0,0000	0,0000	0,0000	0,0000	0.0004

The difference of the proposed presentation of solution (7), (8) from other known solutions is that the convergence of series may be easily improved not only along the sides of the prism but also inside it. The results of computation of stresses at the distance of one twentieth of prism length from the sides are presented in table 5 and table 6. The magnitude of stresses may be explained by the proximity to the eigenfrequency since frequency $\Omega = 1,47$ differs from the one specified in [2] by less than 1%. As it can be seen from these tables the relative error of stresses is greater than 0,001 reaching 0,15 for the stress σ_{xx} at the prism's side. Although the values without convergence improvement may be computed more accurately by increasing the number of equations, the closer the stresses are evaluated to the sides the more equations are required.

Table 5. Stress σ_{xx} along x=0.9a with P=21.

y/b	0,0	0,2	0,4	0,6	0,8	1,0
Without convergence improvement	2,8901	2,7831	2,4642	1,9409	1,2402	0,7544
With convergence improvement	2,8979	2,7909	2,4717	1,9474	1,2412	0,6449

IOP Conf. Series: Journal of Physics: Conf. Series **919** (2017) 012019 doi:10.1088/1742-6596/919/1/012019

Table 6. Stress σ_{yy} along y=0.9b with P=21.

x/a	0,0	0,2	0,4	0,6	0,8	1,0
Without convergence improvement	9,87883	10,2297	11,2212	12,6790	14,3358	15,6916
With convergence improvement	9,8522	10,2032	11,1955	12,6556	14,3248	16.1961

5. Conclusion

In this paper the new presentation of a solution for steady-state vibrations of rectangular orthotropic prism is described. Corresponding infinite system of linear algebraic equations was obtained by the superposition method. A countable set of precise eigenfrequencies and elementary eigenforms was found. The identities were created which make it possible to find the intervals where the infinite system is regular and to discover an asymptotic behavior of variables. That asymptotic law allows improving convergence of the infinite series and finding the analytical sums for all the infinite series in the solution of the problem within the prism and on its faces. The numerical comparison of stresses on the faces of the prism with and without convergence improvement is performed.

References

[1] Grinchenko V T and Meleshko V V 1981 *Harmonic Vibrations and Waves in Elastic Bodies* (Kyev: Naukova Dumka) p 284

[2] Papkov S O 2013 Steady-state forced vibrations of a rectangular orthotropic plate *J. Math. Sci.* 6 **172** 691-702

[3] Gorman D J 2009 Accurate in-plane free vibration analysis of rectangular orthotropic plates *J. Sound and Vib.* **323** 426-43

[4] Gorman D J 2004 Free in-plane vibration analysis of rectangular plates by the method of superposition *J. Sound and Vib.* **272** 831-51

[5] Grinchenko V T 1978 *Equilibrium and Steady-state Oscillations of Elastic Bodies of Finite Dimensions* (Kyev: Naukova Dumka) p 264

[6] Timoshenko S, Woinowsky-Krieger S 1959 *Theory of Plates and Shells* (New York: McGraw-Hill) p 580

[7] Leissa A W 1969 *Vibration of Plates* (NASA SP-160) p 353

[8] Azin A V, Maritskiy N N, Ponomarev S A and Ponomarev S V 2012 Development of a test method on mechanical influences of printed circuit board with chips of BGA and CGA cases *Cur. Issues Cont. Mec. Cel. Mech.* 58-61

[9] Lyashko A D and Chekhov V N 2015 On regularity of infinite systems for steady-state oscillations of orthotropic rectangular prisms *Dyn. Sys.* 1-2 **33** 113-24

[10] Kantorovich L V and Krylov V I 1964 *Approximate Methods of Higher Analysis* (Groningen) p 685

IOP Conf. Series: Journal of Physics: Conf. Series **919** (2017) 012020 doi:10.1088/1742-6596/919/1/012020

Permeability of nanonet structures constructed on the basis of carbon tubes

M A Bubenchikov[1], T I Nikipelova[2], V B Tsyrenova[2] and A S Chelnokova[1,2]

[1] Mechanics and Mathematics Department, National Research Tomsk State University, 36 Lenin ave., Tomsk, Russia 634050
[2] Geometry Department, Institute of Mathematics and Computer Science, Buryat State University, 24a Smolin st., Ulan-Ude, Russia 670000

E-mail: michael121@mail.ru

Abstract. An approximate integration of a potential of paired molecular interactions over the circumferential coordinate is used to find the potential of the interaction between a molecule and an infinite carbon nanotube (CNT). Based on the obtained energy of interaction between the tube and the molecules, local effective radii of the investigated carbon structures with respect to the molecules of the separated gas mixture are found. This makes it possible to calculate permeability of a single-layer, along with a two-layer CNT packing. The conducted research allowed calculating permeability of a nanonet structure as permeability of a two-layer packing of carbon nanotubes for the first time.

1. Introduction

An effective membrane should be both of high-performance and selectivity. Experience shows that productivity of a membrane increases with a decrease in its thickness. Extremely thin membranes of porous graphenes have a thickness of only one atom. From the standpoint of membrane construction on the basis of straight-line carbon wires, a one-layer packing is extremely thin. At the same time, an extremely thin network structure is a two-layer packing of the same rectilinear CNTs. A possibility of theoretical studies on permeability of layers composed of various nanoparticles is provided by the experience which the authors of the present paper have gained in their research in the sphere of nanomechanics. Mechanical resistance of nanoparticles, their centrifugal and gravitational sedimentation, as well as thermophoresis of particles of various shapes, in particular, graphene flakes, was considered in [1-4]. In [5] permeability of porous graphenes was studied from the standpoint of Schrödinger wave dynamics. The mathematical model of undulating graphenes is presented in [6]. The same approach, based on the continual representation of the interaction energy on the basis of the modified *LJ* -potential, was used in [7] to study sorption properties of bodies. In [8], on the basis of the wave permeability model, the authors investigated passage of matter through composite membranes. Moving forward in the direction of theoretical developments, with respect to separation of mixtures, an energy barrier determined by a system of two nanoparticles was found in [9]. Paper [10] is devoted to porous graphenes. Work [11] considers passage of molecules through an open carbon nanotube (CNT). Article [12] presents a mathematical model for a tunnel constructed of spherical carbon nanoparticles. Modes of low-energy interaction of molecules with an open CNT are discussed in [13].

In [14] the subject of research is wave permeability of a layer of compacted spherical particles. Separation of methane-helium mixtures by means of a layer of a porous graphene was considered in [15].

2. The «molecule - infinite CNT» interaction potential

There is an exact analytic solution to the problem of the interaction between a molecule and an infinite tube expressed in terms of hypergeometric functions [16, 17]. In this paper, having carried out an approximate integration with respect to the circumferential coordinate, a compact expression is obtained in the class of functions of the basic potential which provides, like the *LJ*-potential itself, simplicity of computations in a set of statistical data. At the same time, we get a brilliant agreement with the exact solution [16, 17]. There is a basic work on the interaction of surface carbon nanoparticles [18], however, the interaction potential «molecule - fullerene particle» was obtained by V.Y. Rudyak earlier [19]. The latest works by this author relate to physics of the condensed state [20, 21]. Returning to the problems of gas components passage we will briefly describe the scheme for constructing an approximate solution in the case of an infinite nanotube.

The interaction of a test molecule with structure atoms constituting the rectilinear carbon nanotube (CNT) will be considered on the basis of the Lennard-Jones intermolecular interaction potential:

$$U(r)=\frac{C_{12}}{r^{12}}-\frac{C_6}{r^6}, \tag{1}$$

where $C_{12} = 4\varepsilon\sigma^{12}$; $C_6 = 4\varepsilon\sigma^6$; ε, σ are the parameters of the *LJ*-potential.

Despite the fact that interatomic distances in surface carbon structures are 0.141 nm, while their intrinsic transverse size is of the order of 1 nm, the approach of continuous integration in determining the integral effect of nanoparticles has been used since the 1980s [22]. Following the notation introduced by V.Y. Rudyak [19] we can write down the effect on the test molecule from an infinite continuous cylinder by which we mean the surface on which the carbon atoms in the CNT are located:

$$\Phi_S(r)=\sum_{m=12,6}\frac{C_m a}{S_1}\int_0^{2\pi}\left(\int_{-\infty}^{\infty}\left(\rho^2+z^2\right)^{-\frac{m}{2}}dz\right)d\varphi. \tag{2}$$

Here $\rho^2 = a^2 + r^2 - 2ar\cos\varphi$; $ad\varphi dz$ is the area element; S_1 is the area per one molecule (atom) of the structure. The sum consists of two terms, the second term being taken with the minus sign. The integral in parentheses (2) can be taken using the differentiation technology in terms of a parameter or with the help of a package that provides symbolic calculations, for example, «Maple». As a result, we get

$$\int_{-\infty}^{\infty}\left(\rho^2+z^2\right)^{-\frac{m}{2}}dz=b_m\rho^{-m+1},\quad (m=12,6), \tag{3}$$

where $b_{12}=\frac{63}{256}\pi$, $b_6=\frac{3}{8}\pi$.

Then, instead of (12), we can write

$$\Phi_S(r)=\sum_{m=12,6}\frac{b_m C_m}{S_1}\int_0^{2\pi}\rho^{-m+1}d\varphi. \tag{4}$$

Now in the parentheses is the integral over φ from 0 to 2π. This integral contains an embedded cosine function, because ρ depends on $\cos\varphi$. We note here that further application of symbolic calculations leads to an immense analytical result which will be inconvenient in future practical work.

We will take an integral with respect to the angular coordinate approximately by dividing the period $\cos\varphi$ into four segments and using the trapezoid method formula to calculate the integral with respect to φ from the power function of the distance ρ which includes the above-mentioned $\cos\varphi$. This formula contains the values of ρ at five points, but due the symmetry of distribution of $\cos\varphi$ with respect to the line $\varphi = \pi$, only three remain.

As a result, we get

$$\int_0^{2\pi} \rho^{-m+1} d\varphi = \frac{\pi}{2}\left[\rho(0)^{-m+1} + 2\rho\left(\frac{\pi}{2}\right)^{-m+1} + \rho(\pi)^{-m+1}\right]. \tag{5}$$

We note that for the distance from the centre of the test molecule to the material point on the surface of the cylinder the following relations are valid:

$$\rho(0) = r - a;\ \rho\left(\frac{\pi}{2}\right) = \left(r^2 + a^2\right)^{\frac{1}{2}};\ \rho(\pi) = r + a. \tag{6}$$

Introducing (5) and (6) into (4) we find:

$$\Phi_S(r) = K_{11}\left[(r-a)^{-11} + 2\left(r^2+a^2\right)^{-11/2} + (r+a)^{-11}\right] -$$
$$- K_5\left[(r-a)^{-5} + 2\left(r^2+a^2\right)^{-5/2} + (r+a)^{-5}\right].$$

Here r is the distance from the centre of the test molecule to the axis of the tube; a is the radius of the nanotube; $K_{11} = \frac{b_{12}C_{12}a}{S_1}\frac{\pi}{2}$; $K_5 = \frac{b_6C_6a}{S_1}\frac{\pi}{2}$; $b_{12} = \frac{63}{256}\pi$; $b_6 = \frac{3}{8}\pi$; $C_{12} = 4\varepsilon\sigma^{12}$; $C_6 = 4\varepsilon\sigma^6$; S_1 is the area per one molecule (atom) of the tube's wall material; ε, σ are the parameters of the *LJ*-potential.

3. Permeability of a single-layer packing of CNTs

We will consider the process of free molecules passage through a gap between two parallel nanowires. Let us introduce the concept of the width of the local permeability zone:

$$h_{np}(v) = h - 2r_e(v), \tag{7}$$

where h is the distance between the axes of the wires; $r_e(v)$ is the effective radius of the wire calculated as half the limiting distance of molecular passage between the axes with a certain velocity v. Then for molecular systems that obey the Maxwell distribution the relative number of molecules passing between the wires is

$$D(h) = \frac{4}{\sqrt{\pi}}\int_0^{\infty} \alpha^{3/2} e^{-\alpha v^2} v^2 \left[\frac{h_{np}(v)}{h}\right] dv. \tag{8}$$

Here $\alpha = \frac{m}{2kT}$; k is the Boltzmann constant; T is the absolute temperature; m is the mass of the free molecule.

If the expression in square brackets is set equal to one, the integral in (8) is the Maxwell distribution function with respect to the velocity v.

The width of the local permeability zone is determined using the following conditions:

$$h_{np}(v) = \begin{cases} 0, \text{ если } h - 2r_e(v) \le 0, \\ h - 2r_e(v), \text{ если } h - 2r_e(v) > 0, \end{cases} \tag{9}$$

after which the permeability coefficient D is calculated by formula (8).

The distribution $D(h)$ is found in the interval from 0.95 to 5.95 nm (in the case of wires of a radius equal to 0.475 nm) with a digitization every 0.75 nm. It is more convenient to plot lgχ on the vertical axis of separation, since very high values of the degree of mixture separation (more than two orders) are recorded in the range from 1.552 to 1.603 nm.

Figures 1-3 present the summarized data of systematic calculations on the determination of the local (corresponding to the given level of velocities) effective radii of the interaction between a standard CNT (r_p = 0.475 nm) and methane molecules and helium atoms.

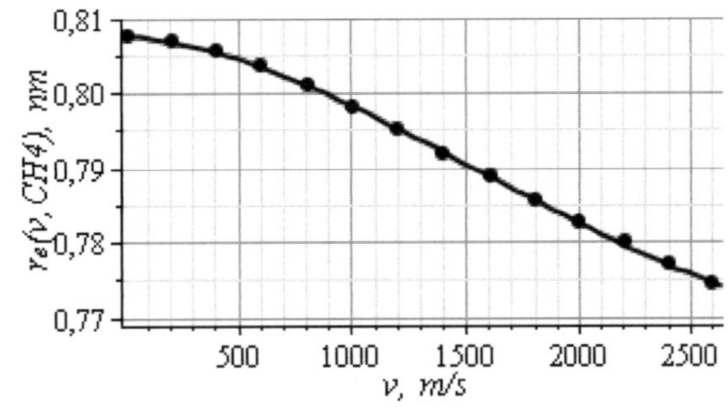

Figure 1. Effective radius of CNT with respect to methane.

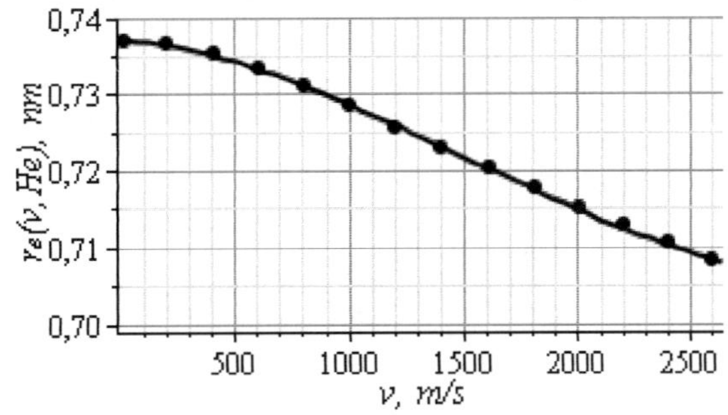

Figure 2. Effective radius of CNT with respect to helium.

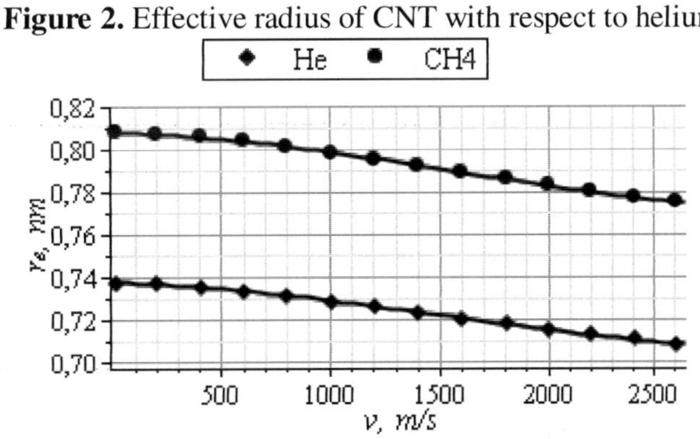

Figure 3. Effective radii of CNTs in relation to methane and helium.

Figures 4 and 5 show calculation results of permeability and the degree of separation of a methane-helium mixture for a single-layer packing of parallel tubes. The generalizing data on the characteristics of the membrane are presented in Table 1.

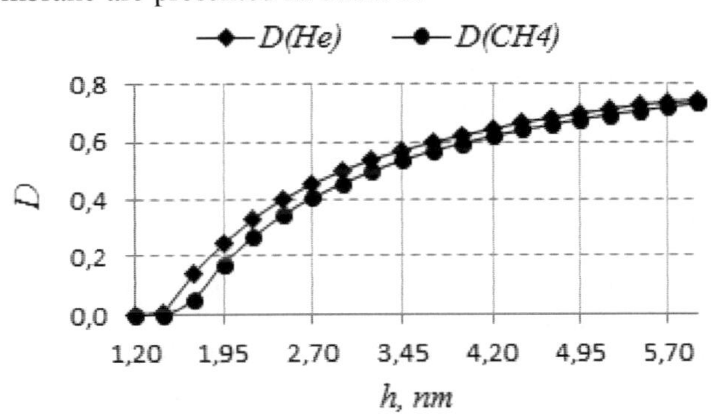

Figure 4. Coefficients of helium and methane passage depending on distance between axes of standard CNTs. Single-layer packing of CNTs.

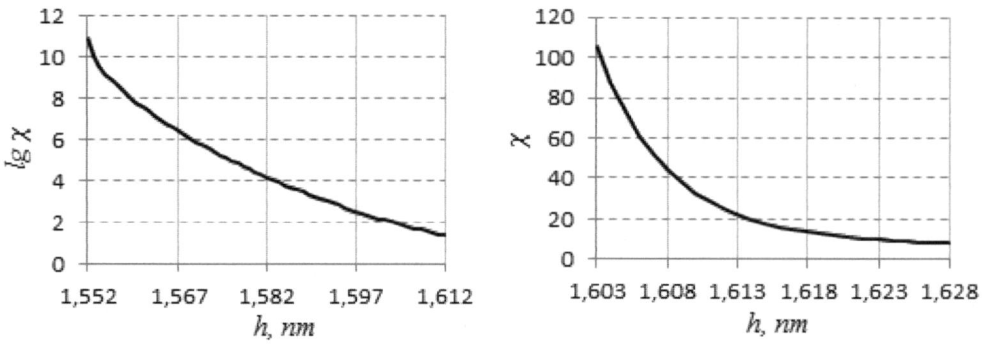

Figure 5. Degree of separation of methane-helium mixture. Single-layer packing of CNTs .

Table 1. Performance of membrane of parallel CNTs (single-layer packing).

Distance between axes of wires, nm	Performance of membrane	Degree of mixture separation
[1,419; 1,552]	Low	Very high
[1,552; 1,603]	Good	Above 100 units
[1,603; 1,755]	High	From 100 to 2 units

The local effective radii of these carbon structures calculated by the method of convergence of two parallel wires have the following analytical approximations (the result is expressed in nm):

$$r_e(v, CH_4) = a_0 v^3 - a_1 v^2 - a_2 v + a_3, \quad (10)$$

$$r_e(v, He) = b_0 v^3 - b_1 v^2 - b_2 v + b_3, \quad (11)$$

where $a_0 = 1.93154 \cdot 10^{-12}, a_1 = 9.09157 \cdot 10^{-9}, a_2 = 2.09756 \cdot 10^{-6}, a_3 = 0.8076,$

$b_0 = 1.71516 \cdot 10^{-12}, b_1 = 7.73388 \cdot 10^{-9}, b_2 = 2.55181 \cdot 10^{-6}, b_3 = 0.73721.$

The limiting distance of approach can be determined from the distribution of the energy of interaction between two identical parallel tubes. Such a distribution was obtained in [20].

Calculations of permeability show that for h varying from 1.419 to 1.552 nm (wider than the limiting zone for carbon) there is a "dead zone" which is characterized by the fact that neither methane molecules nor helium atoms pass between the wires. When analyzing permeability of CNT packings, it should be taken into account that the limiting distance of approach for these carbon structures is 0.32 nm.

4. Permeability of a two-layer packing of tubes

Let us denote by D_1 the value of the fraction of molecules passing through the first stage, and by D_2 - the fraction of those passing through the second stage of the separation system composed of two layers of tubes. From the previous we can write:

$$D_1 = D. \quad (12)$$

where D is calculated by formula (8).

Now we will consider the second step. Passage through the second layer can be considered as a complex event the probability of which is determined by the product of the transmission probabilities of each of the layers. In addition, multiple reflections of molecules trapped between the two layers should be taken into account, as well. As a result, eventually, they are equally likely to come out in one direction or in the other moving along the normal to the layers of the packing.

Since the probability of passage through each of the following for the first layer is equal to D, the second layer will be passed by the following part of molecules incident on the surface of the first layer:

$$D_2 = D_1 D' + \frac{D'}{2}. \tag{13}$$

The first term determines the contribution due to the direct passage through the double layer, the second - the contribution of the re-reflections. The value of D' in the last relation takes into account the increased intensity of molecular motion in the space between the packings which is determined by the following circumstance. Molecules moving through the permeability zone between the two wires pass through the potential well where they acquire additional velocity due to the action of the van der Waals forces from the second row of the packing. Therefore, when free particles (molecules) enter the mode of re-reflections between the two layers of the packing, they move with a velocity higher than that which they had when approaching the carbon structure. Because of this, passage through the second layer is not equivalent to passage through the first layer. The average value of the increase in velocity is determined by the average value of the depth of the potential well.

Being located between the layers, molecules receive an increase in speed of the following order:

$$\Delta v = \frac{2\Delta U}{m(v + v')}. \tag{14}$$

Here ΔU is the average depth of the potential well; v is the velocity of the molecule as it approaches the first layer; v' is the rate of re-reflections.

Remembering that

$$v' = v + \Delta v \tag{15}$$

from (14) we find:

$$\Delta v = \sqrt{v^2 + \frac{2}{m}\Delta U} - v. \tag{16}$$

The last relation shows that, in the general case, molecules located between the layers no longer obey the Maxwell distribution. However, if the depth of the well is large, the value of the increase in velocity is no longer dependent on the velocity itself:

$$\Delta v \approx \sqrt{\frac{2}{m}\Delta U}. \tag{17}$$

It should be noted that at distances of the order of the nanotubes limiting approach (this distance lies between 0.3 nm for fullerenes and 0.34 nm for graphene sheets in graphite) we deal with exactly this kind of situation. In this case, the nature of the Maxwell distribution does not change, but the shift along the velocity axis equal to Δv is fixed to the right. Therefore, D', which is included in formula (13), is defined as follows:

$$D'(h) = \frac{4}{\sqrt{\pi}} \int_{\Delta v}^{\infty} \alpha^{3/2} e^{-\alpha v^2} v^2 \left[\frac{h_{np}(v)}{h}\right] dv, \tag{18}$$

where the lower limit of integration Δv is determined from (17) and ΔU is found from calculations of the energy distribution of the interaction between the molecule and all the surrounding tubes.

Thus, those molecules which accelerated in the potential well and got engaged in the process of re-reflection, pass through the second layer better compared with the moleculeswhich are in front of the first layer of the CNT packing.

Calculations were carried out for the case $D' = D$. It was found that the range of the packing step which provides a high degree of separation is somewhat wider than in the single-layer system, and the degree of separation is higher (Table 2).

Table 2. Performance of membrane of parallel CNTs (two-layer packing).

IOP Conf. Series: Journal of Physics: Conf. Series **919** (2017) 012020 doi:10.1088/1742-6596/919/1/012020

Distance between axes of wires, nm	Performance of membrane	Degree of mixture separation
[1,419; 1,552]	Low	Very high
[1,552; 1,604]	Good	Above 100 units
[1,604; 1,802]	High	From 100 to 2 units

In figures 6 and 7 are the distributions of the components relative permeability and the degree of separation of a methane-helium mixture by a two-layer packing of nanotubes.

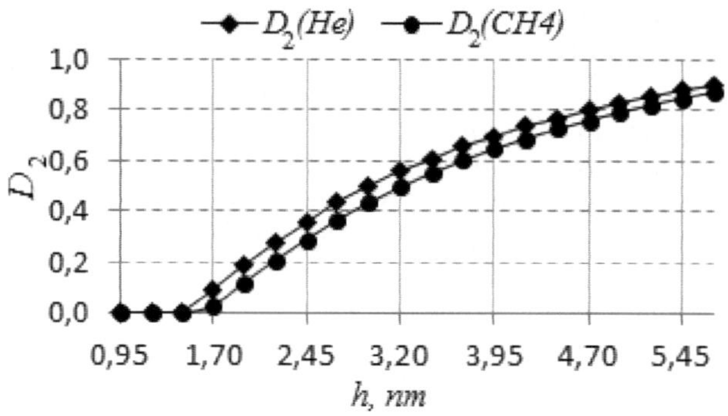

Figure 6. Coefficients of passage of molecules through two-layer packing of CNTs.

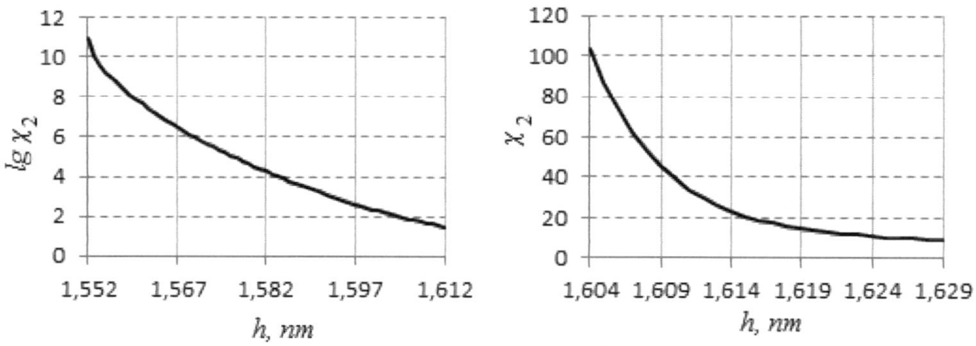

Figure 7. Degree of separation of He/CH_4. Two-layer packing of CNTs.

5. Conclusion

When analyzing mass transfer through a packing of CNTs, the method of local molecular transmission zones proves to be effective. The method consists in finding the effective radii of CNTs corresponding to each particular velocity of a molecule approaching a wire. On the basis of these data, it is possible to determine the relative width of the local molecular transmission zone $\frac{h_{np}(v)}{h}$. Further, this distribution is used in conjunction with the Maxwellian velocity distribution to find the total number of molecules passing through a single-layer packing. The probability of passing through a two-layer packing of tubes is made up of the probability of direct passage through two successive barriers and a fraction of molecules passing through the entire separation system as a result of re-reflections within this system. The found permeability of a double layer of a straight tubes packing simulating a network structure is used to determine the degree of separation of a binary gas mixture.

References

[1] Potekaev A I, Bubenchikov A M and Bubenchikov M A 2013 *Russ. Phys. J.* **55** 1434. DOI: 10.1007 / s11182-013-9977-8

[2] Bubenchikov M A, Potekaev A I and Bubenchikov A M 2013 *Russ. Phys. J.* **56** 341. DOI: 10.1007 / s11182-013-0038-0
[3] Bubenchikov M A, Potekaev A I and Bubenchikov A M 2013 *Russ. Phys. J.* **56** 785. DOI: 10.1007 / s11182-013-0100-y
[4] Bubenchikov A M, Bubenchikov M A, Potekaev A I, Maslov A S, Ovcharenko V V and Usenko O V 2014 *Russ. Phys. J.* **57** 956. DOI: 10.1007 / s11182-014-0330-7
[5] Bubenchikov A M, Bubenchikov M A, Potekaev A I, Libin E E and Khudobina Y P 2014 *Russ. Phys. J.* **57** 1126. DOI: 10.1007 / s11182-014-0353-0
[6] Bubenchikov A M, Bubenchikov M A, Potekaev A I, Libin E E and Khudobina Y P 2015 *Russ. Phys. J.* 58 882. DOI: 10.1007 / s11182-015-0586-6
[7] Bubenchikov A M, Bubenchikov M A, Potekaev A I, Usenko O V, Zhambaa S and Kulagina V V 2016 *Russ. Phys. J.* **58** 1711. DOI: 10.1007 / s11182-016-0706-y
[8] Bubenchikov A M, Bubenchikov M A, Potekaev A I, Libin E E, Khudobina Y P and Kulagina V V 2017 *Russ. Phys. J.* **60** 140. DOI: 10.1007 / s11182-017-1052-4
[9] Bubenchikov A M, Bubenchikov M A, Potekayev A I, Maslov A S, Ovcharenko V V and Usenko O V 2014 *«Izvestiya Vuzov». Physics* **57** 35
[10] Bubenchikov A M, Bubenchikov M A, Potekayev A I, Maslov A S, Ovcharenko V V and Usenko O V 2014 *«Izvestiya Vuzov». Physics* **57** 6
[11] Bubenchikov A M, Bubenchikov M A, Potekayev A I, Klykov I I, Korobitsyn V A and Usenko O V 2014 *«Izvestiya Vuzov». Physics* **57** 15
[12] Bubenchikov A M, Bubenchikov M A, Potekayev A I, Usenko O V and Sherstobitov A A 2014 *Tomsk State University Journal of Mathematics and Mechanics* **5** 69
[13] Bubenchikov M A, Bubenchikov M A, Usenko O V and Tarasov E A 2016 *Tomsk State University Journal of Mathematics and Mechanics* **3** 58
[14] Bubenchikov A M, Bubenchikov M A, Poteryaeva V A and Libin E E 2016 *Tomsk State University Journal of Mathematics and Mechanics* **3** 51
[15] Bubenchikov A M, Bubenchikov M A, Tarasov E A, Usenko O V and Chelnokova A S 2017 *Tomsk State University Journal of Mathematics and Mechanics* **45** 80
[16] Rekhviashvili S S and Kishtikova E V 2010 *Physics and Chemistry of Surfaces and Materials Protection* **46** 51
[17] Rekhviashvili S S 2014 *Dimensional phenomena in physics of condensed states and nanotechnologies* (Nalchik: Kabardino-Balkar Scientific Centre of RAS) p 250
[18] Girifalco L A, Hodak M and Lee R S 2000 *Physical Review B* **62** 13104. DOI: 10.1103/PhysRevB.62.13104
[19] Rudyak V Ya and Krasnolutsky S L 1998 *Statistical mechanics of heterogeneous media. IV. The potential of interaction between molecules and dispersed particles* (Novosibirsk: NSACU) p 46 (Preprint of NSACU 3(13)-98)
[20] Rudyak V Ya and Krasnolutskii S L 2016 *Atmospheric and Oceanic Optics* **29** 512. DOI: 10.1134 / S1024856016060142
[21] Rudyak V Ya, Minakov A V and Pryazhnikov M I 2017 *Technical Physics Letters* **43** 23. DOI: 10.1134 / S106378501612021X
[22] Goodman F and Wahman G 1980 *Dynamics of gas scattering by a surface* (Moscow: «Mir») p 426

IOP Conf. Series: Journal of Physics: Conf. Series **919** (2017) 012021 doi:10.1088/1742-6596/919/1/012021

The boundary element method applied to 3D magneto-electro-elastic dynamic problems

L A Igumnov[1], I P Markov[1] and Iu A Kuznetsov[2]

[1] Research Institute for Mechanics, National Research Lobachevsky State University of Nizhni Novgorod, 23, bldg. 6, Prospekt Gagarina (Gagarin Avenue), Nizhny Novgorod, 603950, Russia
[2] National Research Lobachevsky State University of Nizhni Novgorod, 23, Prospekt Gagarina (Gagarin Avenue), Nizhny Novgorod, 603950, Russia

E-mail: teanku@gmail.com

Abstract. Due to the coupling properties, the magneto-electro-elastic materials possess a wide number of applications. They exhibit general anisotropic behaviour. Three-dimensional transient analyses of magneto-electro-elastic solids can hardly be found in the literature. 3D direct boundary element formulation based on the weakly-singular boundary integral equations in Laplace domain is presented in this work for solving dynamic linear magneto-electro-elastic problems. Integral expressions of the three-dimensional fundamental solutions are employed. Spatial discretization is based on a collocation method with mixed boundary elements. Convolution quadrature method is used as a numerical inverse Laplace transform scheme to obtain time domain solutions. Numerical examples are provided to illustrate the capability of the proposed approach to treat highly dynamic problems.

1. Introduction

Magneto-electro-elastic (MEE) materials that have the ability of converting the energy from one type to another (between mechanical, electric and magnetic energies) possess a wide range of technical applications. They are used in various smart structural devices like transducers, sensors, actuators and also in acoustic devices, ultrasonic imaging, lasers, hydrophones. MEE materials are receiving increasing attention because of providing superior properties compared to the other smart materials due to their magnetic-electric-mechanical coupling effect. However, significant multi field coupling makes static and dynamic problems of linear magneto-electro-elasticity harder to solve.

Numerical simulation plays a very important role in the design and construction of engineering structures and their elements, especially when composite multi-field materials are concerned, because they usually brittle and anisotropic that makes the mathematical modeling of their behaviour more complex. Therefore, it is important to develop a suitable and reliable numerical tool for accurate analysis of static and transient three-dimensional linear magneto-electro-elastic problems with general structural configurations under various loading situations.

Analytical, semi analytical approaches and various numerical methods have been presented for static, time-harmonic and transient dynamic two-dimensional and three-dimensional homogenous and multi-domain problems of linear magneto-electro-elasticity [1-10]. Relatively little work has been done for numerical modeling of dynamic behaviour of magneto-electro-elastic solids and structures with Boundary Element Method (BEM). The key features of the BEM are as follows: the dimensionality reduction so that only boundary discretization is required, higher accuracy due to semi-analytical nature of the method and easier application for infinite and semi-infinite domain problems. However, while BEM offers certain advantages over domain-based methods like Finite Element Method or Finite Differences Method, the corresponding fundamental solutions needed to be known.

In this paper, we apply Laplace domain direct boundary element approach based on the weakly-singular boundary integral equations to three-dimensional dynamic analysis of anisotropic magneto-electro-elastic solids. Three dimensional static fundamental solutions (Green's functions) are represented in an implicit form as an integral over a unit circumference [11, 12]. Dynamic Green's functions are expressed as surface integrals over a unit sphere [13]. For spatial discretization of the displacement boundary integral equations we employ a collocation method with mixed boundary elements. Dynamic parts of the fundamental solutions are interpolated over the boundary elements to reduce overall computation time [14]. Time domain solutions are obtained by a Convolution Quadrature Method (CQM) which is represented as a numerical Laplace transform inversion routine [15]. Transient dynamic numerical examples are provided to show the reliability and versatility of the proposed boundary element formulation.

2. Problem Statement

Consider a fully anisotropic three-dimensional magneto-electro-elastic solid $\Omega \subset R^3$ which can be homogeneous or may contain finite-sized imperfections like cavities. The boundary of the solid is denoted by $\Gamma = \partial\Omega$. Under the quasi-static assumption for the electric and the magnetic fields, implying zero initial conditions and in the absence of body forces, free electrical charges and the magnetic induction sources, the Laplace transforms of dynamic equilibrium equations are given as follows:

$$C_{ijkl}\bar{U}_{k,il}(\mathbf{x},s) = \rho s^2 \delta^*_{jk} \bar{U}_k(\mathbf{x},s), \ \ \mathbf{x} \in \Omega, \ i,l = \overline{1,3}, \ \ j,k = \overline{1,5}, \tag{1}$$

$$\delta^*_{jk} = \begin{cases} \delta_{jk}, & j,k = \overline{1,3}, \\ 0, & \text{otherwise}, \end{cases}$$

where s is the parameter of the Laplace transform, ρ is the mass density. Here the standard contracted notation has been used for the generalized displacements $\bar{U}_k$ and the generalized elasticity tensor C_{ijkl}:

$$\bar{U}_k = \begin{cases} \bar{u}_k, & k = \overline{1,3}, \\ \bar{\varphi}, & k = 4, \\ \bar{\psi}, & k = 5, \end{cases} \tag{2}$$

$$C_{ijkl} = \begin{cases} C^E_{ijkl}, & i,j,k,l = \overline{1,3}, \\ e_{lij}, & i,l,j = \overline{1,3}, k = 4, \\ e_{ikl}, & i,l,k = \overline{1,3}, \ j = 4, \\ q_{lij} & i,l,j = \overline{1,3}, k = 5, \\ q_{ikl} & i,l,k = \overline{1,3}, \ j = 5, \\ -\lambda_{il} & i,l = \overline{1,3}, \ j = 4, k = 5 \text{ or } k = 4, \ j = 5, \\ -\varepsilon_{il} & i,l = \overline{1,3}, \ j,k = 4, \\ -\mu_{il} & i,l = 1,3, \ j,k = 5, \end{cases} \tag{3}$$

where $\bar{u}_k$, $\bar{\varphi}$ and $\bar{\psi}$ stand for the elastic displacements, electric potential, and magnetic potential, respectively; C^E_{ijkl}, e_{lij} and q_{lij} denote the elastic, piezoelectric and piezomagnetic tensors, respectively; λ_{il}, ε_{il} and μ_{il} indicate the magnetoelectric coefficients, dielectric permittivity and magnetic permittivity tensors, respectively. The following symmetry relations are satisfied:

$$C_{ijkl} = C_{lkji},\ e_{lij} = e_{lji},\ q_{lij} = q_{lji},\ \lambda_{il} = \lambda_{li},\ \varepsilon_{il} = \varepsilon_{li},\ \mu_{il} = \mu_{li}. \tag{4}$$

The generalized boundary conditions on Γ are expressed as

$$\bar{U}_k(\mathbf{x},s) = \bar{U}_k^*(\mathbf{x},s),\ \ \mathbf{x}\in\Gamma_U, \tag{5}$$

$$\bar{T}_k(\mathbf{x},s) = \bar{T}_k^*(\mathbf{x},s),\ \ \mathbf{x}\in\Gamma_T, \tag{6}$$

with

$$\bar{T}_j = \begin{cases} \bar{t}_j = \bar{\sigma}_{jk}n_k, & j=\overline{1,3}, \\ \bar{D}_n = \bar{D}_k n_k, & j=4, \\ \bar{B}_n = \bar{B}_k n_k & j=5, \end{cases} \tag{7}$$

where $\bar{\sigma}_{ij}$, $\bar{D}_i$ and $\bar{B}_i$ are the Cauchy stress tensor, electric displacement vector and magnetic induction vector, respectively; n_k are the components of the external unit normal vector to the boundary.

3. BEM Formulation

The weakly-singular boundary integral equations follow from the extended Somigliana equation and take the form

$$\int_\Gamma \left[\bar{U}_k(\mathbf{y},s)\bar{h}_{jk}(\mathbf{y},\mathbf{x},s) - \bar{U}_k(\mathbf{x},s)h_{jk}^S(\mathbf{y},\mathbf{x})\right]d\Gamma(\mathbf{y}) - \int_\Gamma \bar{T}_k(\mathbf{y},s)\bar{g}_{jk}(\mathbf{y},\mathbf{x},s)d\Gamma(\mathbf{y}) = 0,\ \ \mathbf{x}\in\Gamma, \tag{8}$$

where no strongly singular integrals present; $\bar{g}_{jk}(\mathbf{y},\mathbf{x},s)$ and $\bar{h}_{jk}(\mathbf{y},\mathbf{x},s)$ are the Laplace transformed dynamic fundamental solutions for the generalized displacements and tractions at the field point $\mathbf{y}$ due to a unit source applied at the collocation point $\mathbf{x}$; $h_{jk}^S(\mathbf{y},\mathbf{x})$ is the static part of $\bar{h}_{jk}(\mathbf{y},\mathbf{x},s)$.

The generalized traction fundamental solution is obtained using the relation:

$$\bar{h}_{jp}(\mathbf{y},\mathbf{x},s) = C_{ijkl}\bar{g}_{kp,l}(\mathbf{y},\mathbf{x},s)n_i(\mathbf{y}),\ \ j,p=\overline{1,5}, \tag{9}$$

where $n_i(\mathbf{y})$ are the components of the unit normal vector to the boundary at the field point $\mathbf{y}$.

The Laplace transformed generalized displacement fundamental solution is represented as a sum of singular and regular parts as follows:

$$\bar{g}_{jp}(\mathbf{y},\mathbf{x},s) = \bar{g}_{jp}(\mathbf{r},s) = g_{jp}^S(\mathbf{r}) + \bar{g}_{jp}^R(\mathbf{r},s),\ \ \mathbf{r}=\mathbf{y}-\mathbf{x},\ \ r=|\mathbf{r}|, \tag{10}$$

where the singular part $g_{jp}^S(\mathbf{r})$ represents the static magneto-electro-elastic fundamental solution.

Following the works [11-13] singular and regular terms can be expressed as

$$g_{jp}^S(\mathbf{r}) = \frac{1}{8\pi^2 r}\int_{|\mathbf{d}|=1}\Gamma_{jp}^{-1}(\mathbf{d})dL(\mathbf{d}),\ \ \bar{g}_{jp}^R(\mathbf{r},s) = -\frac{1}{8\pi^2}\int_{\substack{|\mathbf{n}|=1\\ \mathbf{n}\cdot\mathbf{r}>0}}\sum_{m=1}^{Q}\frac{k_m\bar{P}_{jp}^m(\mathbf{n})}{\rho c_m^2}e^{-k_m|\mathbf{n}\cdot\mathbf{r}|}dS(\mathbf{n}), \tag{11}$$

$$\bar{P}_{jp}^m(\mathbf{n}) = \begin{cases} E_{jp}^m / E_{kk}^m, & j,p=\overline{1,3}, \\ \alpha_l^j E_{lp}^m / E_{kk}^m, & j=\overline{4,5},\ p=\overline{1,3},\ i,k,l=\overline{1,3}, \\ \alpha_l^j\alpha_k^p E_{lk}^m / E_{kk}^m, & j=p=\overline{4,5}, \end{cases} \tag{12}$$

$$E^m_{il}(\mathbf{n}) = \operatorname{adj}\left(Z_{il} - \rho c_m^2 \delta_{il}\right),\ Z_{ik}(\mathbf{n}) = \Gamma_{ik} + \alpha_k^4 \Gamma_{i4} + \alpha_k^5 \Gamma_{i5},\ \Gamma_{jp}(\mathbf{a}) = C_{ijkl} a_k a_l, \tag{13}$$

$$\alpha_k^4(\mathbf{n}) = \frac{\Gamma_{4k}\Gamma_{55} - \Gamma_{45}\Gamma_{5k}}{\Gamma_{45}\Gamma_{54} - \Gamma_{44}\Gamma_{55}},\ \alpha_k^5(\mathbf{n}) = \frac{\Gamma_{44}\Gamma_{5k} - \Gamma_{4k}\Gamma_{54}}{\Gamma_{45}\Gamma_{54} - \Gamma_{44}\Gamma_{55}},\ c_m = \sqrt{\lambda_m / \rho},\ k_m = s/c_m, \tag{14}$$

where λ_m are the eigenvalues of the $Z_{il}(\mathbf{n})$, Q is the number of distinct eigenvalues, c_m are the phase velocities of the elastic waves and k_m are the corresponding wave numbers.

The domains of integration for the dynamic and static parts of the fundamental solutions are defined as follows (figure 1):

$$dL\left(\mathbf{d}(\varphi)\right) \in D^S = \{0 \le \varphi \le 2\pi\},\ dS\left(\mathbf{n}(b,\varphi)\right) \in D^R = \{0 \le b \le 1; 0 \le \varphi \le 2\pi\}, \tag{15}$$

$$\mathbf{n}(b,\varphi) = \sqrt{1-b^2}\,\mathbf{d} + b\mathbf{e},\ \mathbf{e} = \mathbf{r}/r,\ \mathbf{e} = \left[e_1, e_2, e_3\right], \tag{16}$$

$$\mathbf{d}(\varphi) = \left[e_2 \cos\varphi + e_1 e_3 \sin\varphi, -e_1 \cos\varphi + e_2 e_3 \sin\varphi, -\left(1-e_3^2\right)\sin\varphi\right] \Big/ \sqrt{1-e_3^2}\,. \tag{17}$$

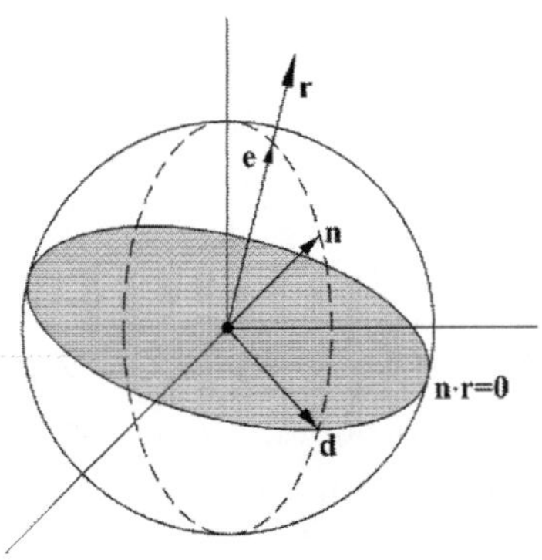

Figure 1. A schematic representation of vectors **r**, **e**, **n** and **d** involved in numerical evaluation of magneto-electro-elastic fundamental solutions.

In this formulation the mixed boundary elements are employed for spatial discretization of equation (8). Boundary Γ is discretized with quadratic quadrilateral elements. In order to correctly model continuous generalized displacements and discontinuous generalized tractions in each boundary element, linear and constant shape functions are used, respectively. For efficient numerical implementation of dynamic fundamental solutions the regular parts and their spatial derivatives are interpolated [14] over a boundary elements using the shape functions of five-node element (figure 2):

$$N_m(\xi_1,\xi_2) = \frac{1}{4}\left(1 + \xi_1^{(m)}\xi_1\right)\left(1 + \xi_2^{(m)}\xi_2\right) - \frac{1}{4}N_5,\ N_5(\xi_1,\xi_2) = \left(1-\xi_1^2\right)\left(1-\xi_2^2\right),\ m = \overline{1,4}, \tag{18}$$

where $-1.0 \le \xi_1, \xi_2 \le 1.0$.

Approximation takes place when a particular boundary element is located relatively far from the collocation point, i.e. when $R/l \ge k$ where R is the distance to the boundary element, l is the average side length of all boundary elements and k is a constant (see scheme at figure 3). We find that $k = 2.0$ allows reducing the computation time by almost 50% while retaining the acceptable accuracy of the results. After enforcing the boundary conditions a standard collocation technique leads to a complex-valued system of linear algebraic equations:

$$\overline{\mathbf{A}}_k \overline{\mathbf{p}}_k = \overline{\mathbf{f}}_k, \tag{19}$$

where $\overline{\mathbf{A}}_k \in \mathrm{C}^{\mathrm{N_{dof}} \times \mathrm{N_{dof}}}$ is the influence matrix, $\overline{\mathbf{f}}_k \in \mathrm{C}^{1 \times \mathrm{N_{dof}}}$, $\overline{\mathbf{p}}_k \in \mathrm{C}^{\mathrm{N_{dof}} \times 1}$ contain all the known and unknown boundary nodal values, respectively; $\mathrm{N_{dof}}$ is the total number of unknowns.

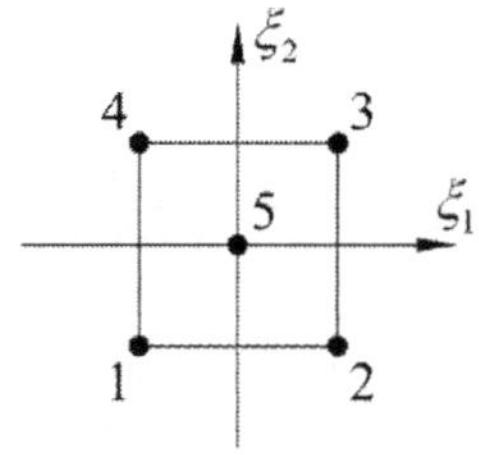

Figure 2. Five-node element.

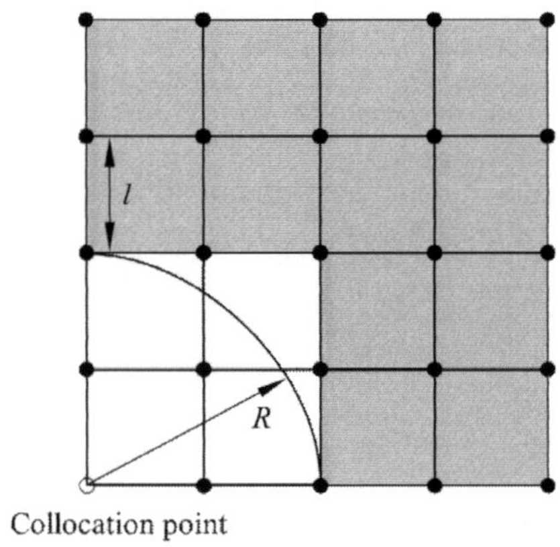

Figure 3. Fundamental solutions and their spatial derivatives are approximated over the boundary elements which are located at the shaded region.

All variables in equation (19) correspond to the specific value of a Laplace transform parameter s_k. In order to obtain a time domain solution we need to assemble and solve system (19) for a set of Laplace transform parameters. Once solutions on a complete set of Laplace transform parameters s_k are obtained, we employ a numerical inversion routine to retrieve time domain solutions.

4. Convolution Quadrature Method for numerical inversion of Laplace transform

With the Convolution Quadrature Method [16, 17] one can approximate a convolution integral

$$f(t) = y(t) * g(t) = \int_0^t y(t-\tau) g(\tau)\,\mathrm{d}\tau, \tag{20}$$

as follows

$$f(n\Delta t) = \sum_{k=1}^{n} \omega_k(\overline{y}) g(k\Delta t), \;\; n = \overline{1, N}, \tag{21}$$

where $\overline{y}$ is the Laplace transform of $y(t)$, Δt is the time step size and N is the total number of time steps so $[0, T]$ is the total time interval with $T = N\Delta t$.

Now let $\overline{f}(s)$ be the Laplace transform of $f(t)$. Then,

$$f(t) = L^{-1}\left[\overline{f}(s)\right], \tag{22}$$

where L^{-1} denotes the inverse Laplace transform operator and s is the Laplace parameter.

We introduce the following auxiliary functions $\overline{F}_1(s) = s\overline{f}(s)$, $\overline{F}_2(s) = 1/s$. By definition,

$$F_1(t) = L^{-1}\left[\overline{F}_1(s)\right], \;\; F_2(t) = L^{-1}\left[\overline{F}_2(s)\right]. \tag{23}$$

The transformed function $\overline{f}(s)$ can be rewritten as

$$\overline{f}(s) = \overline{F}_1(s)\overline{F}_2(s). \tag{24}$$

Therefore, we have

$$f(t) = L^{-1}\left[\overline{f}(s)\right] = L^{-1}\left[\overline{F}_1(s)\overline{F}_2(s)\right] = F_1(t) * F_2(t). \tag{25}$$

We notice that $F_2(t) = H(t)$, where $H(t)$ is the Heaviside function defined by

$$H(t) = \begin{cases} 1, & t \geq 0 \\ 0, & \text{otherwise.} \end{cases} \tag{26}$$

Now we apply the CQM with the Backward Differential Formula of order two [15, 18] as an underlying multi-step method, to numerically approximate the convolution integral in equation (25) and thus obtain $f(t)$:

$$f(0) = 0, \quad f(n\Delta t) = \sum_{k=1}^{n} \omega_k(\overline{F}_1) H(k\Delta t) = \sum_{k=1}^{n} \omega_k(\overline{F}_1), \quad n = \overline{1, N}, \tag{27}$$

$$\omega_k(\overline{F}_1) = \frac{R^{-k}}{2\pi} \int_0^{2\pi} \overline{F}_1(s_p) e^{-ik\phi} d\phi \approx \frac{R^{-k}}{L} \sum_{p=0}^{L-1} \overline{F}_1(s_p) e^{-ik\phi_p} = \frac{R^{-k}}{L} \sum_{p=0}^{L-1} s_p \overline{f}(s_p) e^{-ik\phi_p}, \tag{28}$$

$$s_p = \frac{\gamma(z_p)}{\Delta t}, \quad z_p = Re^{i\phi_p}, \quad \phi_p = 2\pi \tfrac{p}{L}, \quad \gamma(z_p) = 3/2 - 2z_p + z_p^2/2, \tag{29}$$

where the L-point trapezoidal rule with equal steps $2\pi/L$ for numerical evaluation of the weights ω_k, $0 < R < 1$ is a CQM parameter.

5. Numerical examples

Consider a magneto-electro-elastic cube with an edge length of 1 m. At the bottom face $x_3 = 0$ m the following boundary conditions are prescribed: $u_1 = 0$, $u_2 = 0$, $u_3 = 0$, $\phi = 0$, $\psi = 0$; at the upper face $x_3 = 1$ m the cube is subjected to uniaxial and uniform impact loading $t_3 = t_3^* H(t)$, $t_3^* = -100$ Pa with $H(t)$ being the Heaviside function. Two cases are considered: when the cube is homogeneous and when it contains a central cubic cavity with the side length of 0.45 m (figure 4). The rest of the faces are traction free. The cube is made of $BaTiO_3$-$CoFe_2O_4$ with density $\rho = 5550$ kg/m^3 and with the following effective material properties [19]:

$$\mathbf{C}^E = \begin{bmatrix} 213 & 113 & 113 & 0 & 0 & 0 \\ 113 & 213 & 113 & 0 & 0 & 0 \\ 113 & 113 & 207 & 0 & 0 & 0 \\ 0 & 0 & 0 & 49.9 & 0 & 0 \\ 0 & 0 & 0 & 0 & 49.9 & 0 \\ 0 & 0 & 0 & 0 & 0 & 50 \end{bmatrix} \text{GPa}, \quad \mathbf{e} = \begin{bmatrix} 0 & 0 & 0 & 0 & 0.15 & 0 \\ 0 & 0 & 0 & 0.15 & 0 & 0 \\ -2.71 & -2.71 & 8.86 & 0 & 0 & 0 \end{bmatrix} \text{C/m}^2,$$

$$\boldsymbol{\varepsilon} = \begin{bmatrix} 0.24 & 0 & 0 \\ 0 & 0.24 & 0 \\ 0 & 0 & 6.37 \end{bmatrix} \cdot 10^{-9}\,\text{C/V}\cdot\text{m}, \quad \mathbf{q} = \begin{bmatrix} 0 & 0 & 0 & 0 & 185 & 0 \\ 0 & 0 & 0 & 185 & 0 & 0 \\ 222 & 222 & 292 & 0 & 0 & 0 \end{bmatrix} \text{N/A}\cdot\text{m},$$

$$\boldsymbol{\lambda} = \begin{bmatrix} -5.23 & 0 & 0 \\ 0 & -5.23 & 0 \\ 0 & 0 & 2750 \end{bmatrix} \cdot 10^{-12}\,\text{N}\cdot\text{s/V}\cdot\text{C}, \quad \boldsymbol{\mu} = \begin{bmatrix} 2.01 & 0 & 0 \\ 0 & 2.01 & 0 \\ 0 & 0 & 0.839 \end{bmatrix} \cdot 10^{-4}\,\text{N}\cdot\text{s}^2/\text{C}^2.$$

IOP Conf. Series: Journal of Physics: Conf. Series **919** (2017) 012021 doi:10.1088/1742-6596/919/1/012021

The boundary element results for elastic displacements $u_3(t)$, electric potential $\phi(t)$ and magnetic potential $\psi(t)$ at point A (0.5, 0.5, 1.0) m for the homogenous cube and at points A and B (0.5, 0.275, 0.5) m for the cube with a cavity are shown in figures 5-7 and figures 8-10, accordingly. In both cases three different uniform meshes with 216 (mesh 1), 384 (mesh 2), 600 (mesh 3) and 312 (mesh 1), 600 (mesh 2), 984 (mesh 3) elements were used, respectively. The results obtained with the proposed formulation indicate rapid convergence, even for magnetic potential behaviour.

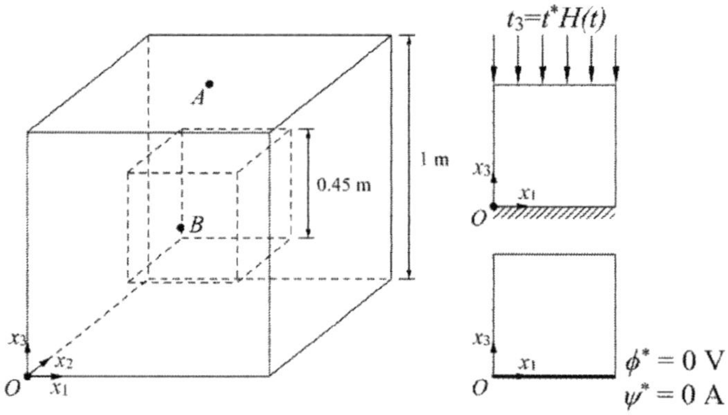

Figure 4. Magneto-electro-elastic cube under impact loading.

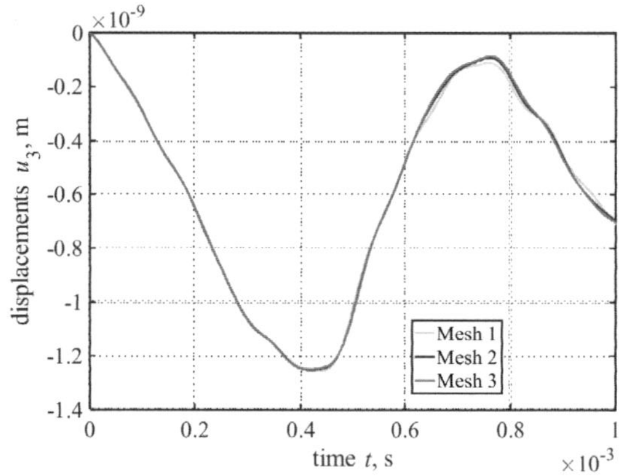

Figure 5. Displacements $u_3(t)$ at point A.

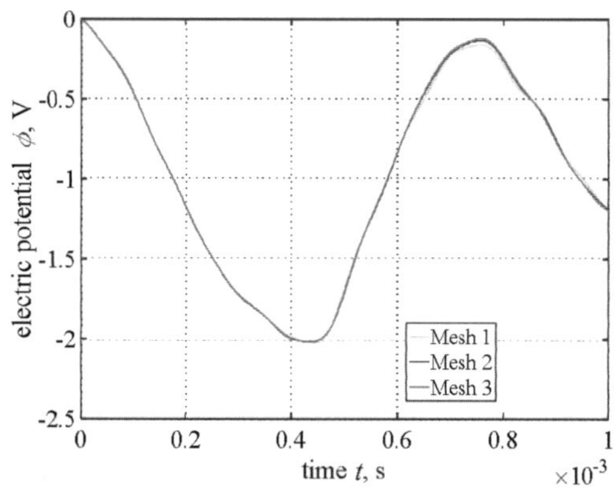

Figure 6. Electric potential $\phi(t)$ at point A.

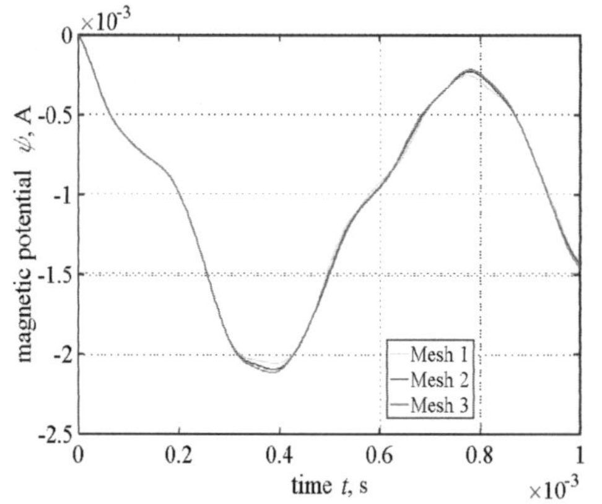

Figure 7. Magnetic potential $\psi(t)$ at point A.

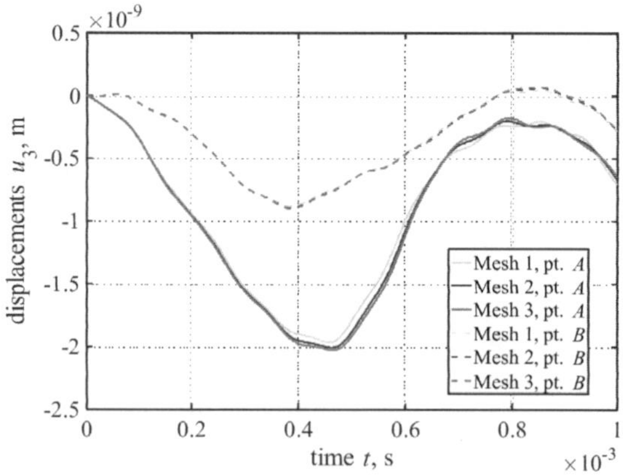

Figure 8. Displacements $u_3(t)$ at pts. A, B.

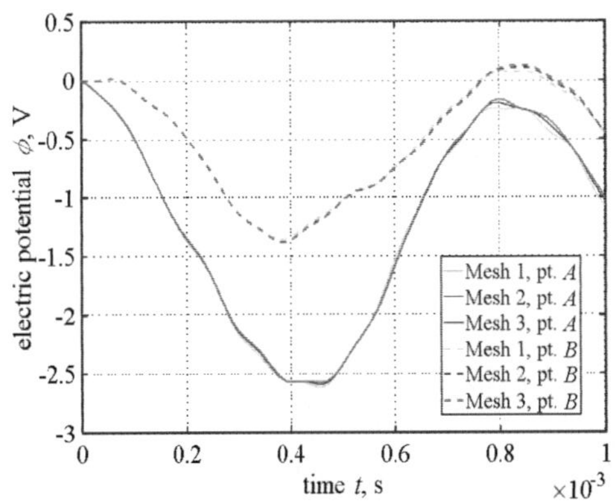

Figure 9. Electric potential $\phi(t)$ at pts. A, B.

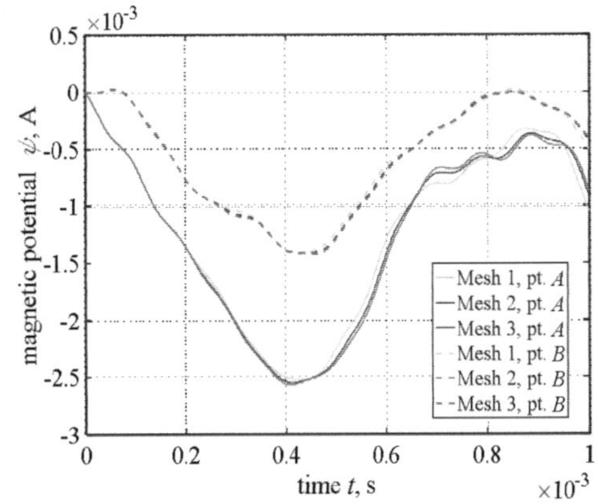

Figure 10. Magnetic potential $\psi(t)$ at pts. A, B.

In the second example a magneto-electro-elastic solid with more complex shape and loading conditions is considered (figure 11). As in the previous example, two cases are examined: when the solid is homogeneous and when a rectangular prismatic cavity is present. The dimensions of the cavity are $0.03 \times 0.03 \times 0.08$ m and the coordinates of its center are (0.025, 0.025, 0.05) m. On the top half of the left face $x_1 = 0.0$ m of the solid the tractions $t_1 = t_1^* H(t)$, $t_1^* = -100$ Pa are applied and on the bottom face $x_3 = 0$ m the following boundary conditions are prescribed: $u_1 = 0$, $u_2 = 0$, $u_3 = 0$, $\phi = 0$, $\psi = 0$. The rest of the faces are traction free. The material is again $BaTiO_3$-$CoFe_2O_4$ with properties defined in the previous example. In figures 12-14 obtained results for the displacements $u_1(t)$, electric potential $\phi(t)$ and magnetic potential $\psi(t)$ at the point A (0.1, 0, 0.1) m for both cases (with and without cavity) are displayed. As it can be observed, the proposed boundary element approach yields stable results for this considerably complex problem.

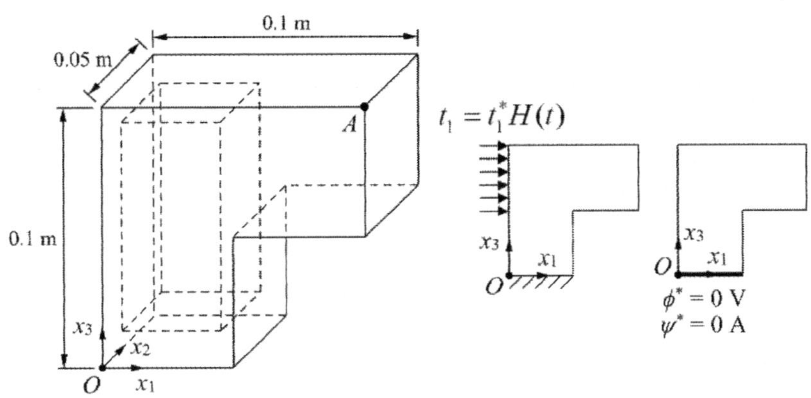

Figure 11. Magneto-electro-elastic solid with complex shape and loading conditions.

6. Conclusions

In this paper we have presented a Laplace domain direct boundary element formulation for transient dynamic problems of three-dimensional linear magneto-electro-elasticity. The Laplace transformed generalized fundamental solution is represented as a sum of singular and regular parts. The dynamic part is expressed as a surface integral over a half of a unit sphere and the singular static part as an integral over a unit circumference. The boundary elements are selected in a way that allows simultaneous modeling of continuous generalized displacements and discontinuous generalized tractions. The spatial discretization is done by a classical collocation scheme. Time domain solutions are retrieved via Convolution Quadrature Method for numerical inversion of Laplace transform. To verify the proposed formulation, a simple numerical example of a magneto electro-elastic cube under uniform uniaxial impact loading is presented. Additionally, the ability to produce stable numerical results was demonstrated on a problem involving a solid with complex shape and under sophisticated magneto-electro-mechanical loading.

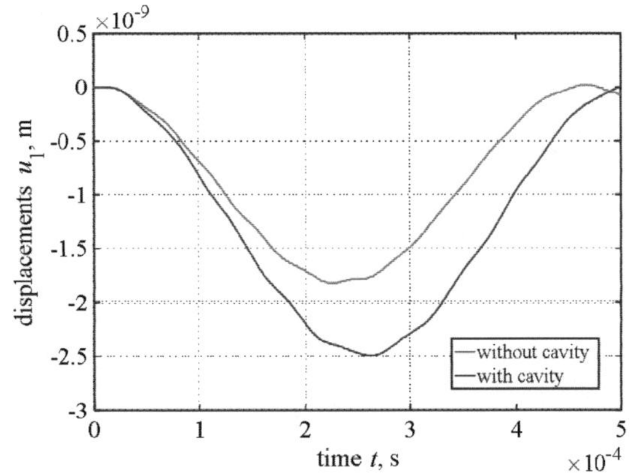

Figure 12. Displacements $u_1(t)$ at point A.

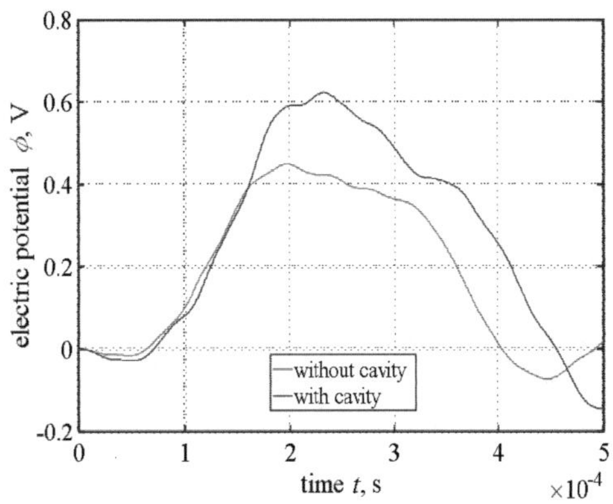

Figure 13. Electric potential $\phi(t)$ at point A.

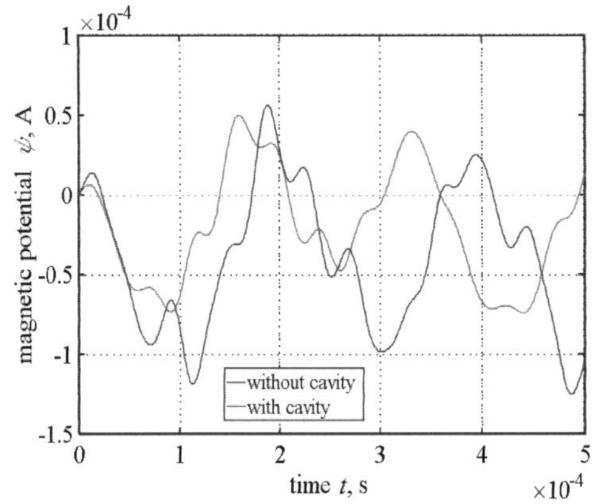

Figure 14. Magnetic potential $\psi(t)$ at point A.

Acknowledgments
The reported study was funded by the Russian Foundation for Basic Research, according to the research project No. 16-38-60097 mol_a_dk.

References

[1] Milazzo A and Orlando C 2012 *Compos. Struct.* **94** 3710-21
[2] Zhu X, Huang Z, Jiang A and Chen W Q 2010 *Eng. Anal. Bound. Elem.* **34** 927-933
[3] Milazzo A, Orlando C and Alaimo A 2009 *Smart Mater. Struct.* **18** 085012
[4] Daga A, Ganesan N and Shankar K 2009 *Int. J. Comput. Methods Eng. Sci. Mech.* **10** 173-185
[5] Daga A, Ganesan N and Shankar K 2009 *J. Intel. Mat. Syst. Str.* **20** 1203-20
[6] Li X C and Yao W A 2006 *Eng. Anal. Bound. Elem.* **30** 709-717
[7] Ding H, Jiang A, Hou P and Chen W 2005 *Eng. Anal. Bound. Elem.* **29** 551-561
[8] Haojiang D and Aimin J 2004 *Compos. Struct.* **82** 1599-1607
[9] Heyliger P R and Pan E 2004 *AIAA J.* **42** 1435-43

[10] Pan E and Heyliger P R 2003 *Int. J. Solids Struct.* **40** 6859-76
[11] Buroni F C and Saez A 2010 *P. R. Soc.* A **466** 515-537
[12] Pan E 2002 *Z Angew. Math. Phys.* **53** 815-838
[13] Rojas-Díaz R, Sáez A, García-Sánchez F and Zhang Ch *Int. J. Solids Struct.* **45** 144-158
[14] Matsumoto T, Tanaka M and Ogawa Y 2003 *Proc. of Second MIT Conference on Computational Fluid and Solid Mechanics* vol 2 (Oxford: Elsevier Science Ltd) pp 2071-73
[15] Igumnov L A, Litvinchuk S Yu, Petrov A N and Ipatov A A 2016 Numerically analytical modeling the dynamics of a prismatic body of two- and three-component materials *Advanced Materials: Manufacturing, Physics, Mechanics and Applications (Springer Proceedings in Physics* vol 175*)* ed I A Parinov *et al* (Cham: Springer International Publishing) chapter 35 pp 505–514
[16] Lubich C 1988 *Numer. Math.* **52** 129-145
[17] Lubich C 1988 *Numer. Math.* **52** 413-425
[18] Schanz M 2001 *Wave Propagation in Viscoelastic and Poroelastic Continua: A Boundary Element Approach* (Springer-Verlag Berlin Heidelberg)
[19] Xue C X and Pan E 2013 *Int. J. Eng. Sci.* **62** 48-55

IOP Conf. Series: Journal of Physics: Conf. Series **919** (2017) 012022 doi:10.1088/1742-6596/919/1/012022

The study of the effect of aluminum powders dispersion on the oxidation and kinetic characteristics

T I Gorbenko[1, 2], M V Gorbenko[2], M P Orlova[1] and S A Volkov[1]

[1] Tomsk State University, 36, Lenina ave., Tomsk, 634050, Russia
[2] Tomsk Polytechnic University, 30, Lenina ave., Tomsk, 634050, Russia

E-mail: gorbenkoti@rambler.ru

Abstract. Differential-scanning calorimetry (DSC) and thermogravimetric analysis (TG) were used to study micro-sized aluminum powder ASD-4 and nano-sized powder Alex. The dependence of the oxidation process on the dispersion of the sample particles is shown. The influence of thermogravimetric conditions on the thermal regime of the process was considered, and its kinetic parameters were determined. Calculations of the activation energy and the pre-exponential factor were carried out.

1. Introduction

Over the past decades, the use of powdered metals is constantly expanding, with the help of which it is possible to improve the energy and performance characteristics of propulsion systems. Aluminum powders have been found wide application as a fuel component in energy condensed systems for various purposes, as well as in space technology [1–3].

The study of the regularities of metal oxidation processes makes it possible to determine the influence degree of both dimensional and structural factors on the reactivity of perspective materials and determine their stability under environmental conditions.

Thermogravimetric study of the influence of phase formation on the kinetics and mechanism of oxidation of aluminum powder doped with calcium was carried out in [4]. The completeness of the conversion of a calcium-doped powder is two or more times higher than that of pure aluminum. The authors [5] investigated the effect of barium on the kinetics of the oxidation of an aluminum-based alloy powder. The comparison of thermograms indicates that the pure aluminum powder up to the melting point oxidizes more vigorously than the alloy. After melting, the alloy powder oxidizes more actively. The maximum exothermal peak of alloy oxidation at 1383 K corresponds to a specific heat release 9026 J/g. Alloying of aluminum with barium increases the completeness and rate of aluminum oxidation. In these studies, the authors did not investigate the effect of aluminum dispersion on the oxidation regularities.

At present nano-sized aluminum powder is widely studied by thermal analyses. But it is also important to know the kinetic characteristics of oxidation of aluminum powder: the activation energy and the pre-exponential factor [6].

Analysis of scientific publications shows that the results of different authors of the kinetic characteristics of metallic powders oxidation differ considerably. It can be assumed that this discrepancy is due to different methods of aluminum powder production, storage conditions, and the content of impurities.

Despite a wide range of studies of metal powders, the relevance of analyzing the effect of the powder particle size on the behavior of physicochemical processes remains.

The present work is devoted to the study of the effect of aluminum powder dispersion on the oxidation and kinetic characteristics using differential scanning calorimetry (DSC) and thermogravimetric analysis (TG).

2. Experimental research

2.1. Samples

Recently, aluminum nanopowders produced by the electrical explosion of wires method became commercialized industrial product. The aluminum nanopowders are widely studied as the promising components for high energy materials of different types. The most important problem of the wide application of aluminum nanopowders is the dependence of their properties from the production conditions.

Aluminum powders of two brands were studied in this work: micro-sized aluminum powder ASD-4, and nano-sized powder Alex obtained by the electric explosion method with average mass diameters of 7.34 μm and 0.18 μm respectively. The specific surface area was 0.5 m^2/g for ASD-4, and 13.9 m^2/g for Alex. The content of active aluminum in the powders of these grades was 98.0 and 85.8% weight, respectively [7].

2.2. Thermal analysis

Today thermal analysis is one of the most common and widely used laboratory techniques. Combustion researchers exploit thermal analysis to characterize both the stability and ignition of many high energy materials.

However, there are important limitations. The rates of thermal analysis measurements are necessarily much lower than the heating rates occurring in most practical situations involving ignition of reactive materials.

To study the effect of dispersion of aluminum powder on oxidation process, the methods of TG and DSC were used. The experiment was carried out in the temperature range from the standard room temperature up to 1000 °C under heating rates of 5, 10 and 20 C/min in the air flow.

When the temperature changes, the processes in the sample, such as melting, restructuring of the crystal structure, evaporation, oxidation, can proceed with a change in the enthalpy.

In the thermal analysis method, the following notation is adopted – endo-effects (heat absorption) assume to be negative, and exo-effects (heat release) to be positive.

The experiment was performed with NETZCH STA 409 PC/PG equipment.

3. Results and discussion

Based on the TG analysis, it was found that when the samples are heated up to ~500°C, the mass increase due to oxidation practically does not occur. Moreover, when the temperature changes from 200 to 500 °C both for Alex and ASD-4, a decrease in mass (up to 1.7%) is observed. The loss of mass may be related to evaporation of the adsorbed water on nanoparticle surfaces or adsorbed gases desorption.

Experimental studies showed that the intensive oxidation of ASD-4 and Alex in air begins below the melting points of aluminum powders and the rate of heating affects the onset of oxidation. Analysis of the curves in figures 1 and 2 showed that for aluminum powder ASD-4 the onset oxidation temperature, depending on the heating rate, varies from 571 to 576 °C and for Alex varies from 550 to 580 °C.

The process of nonisothermic oxidation for aluminum powder may be divided into two stages: low-temperature, due to heterogeneous reactions between the gaseous oxidizer and solid particle core

[3], with a sharp mass gain, (for ASD-4 – 571–650 °C, for Alex –550–625 °C) and high-temperature less intensive stage of oxidation (ASD-4 – 750–830 °C, Alex – 705–740 °C), depending on the heating rate. This two-stage character may be related to any of the following phenomena: the periodic cracking of the oxide shell covering the particle metal core [8]; the separate oxidation of different particle fractions (the finest particles are oxidized before metal core melting, while the larger ones react after melting in the same temperature range as the conventional micron-sized powders [9].

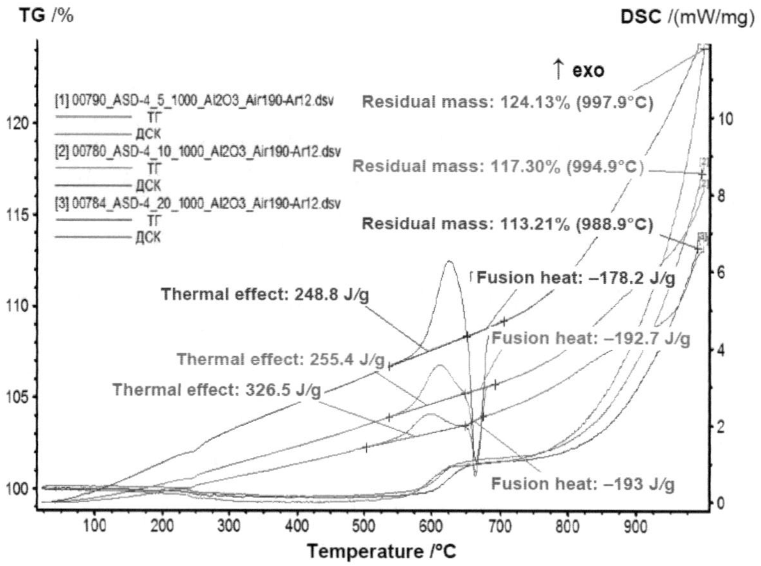

Figure 1. TG and DSC diagrams for ASD-4 at heating rates of 5, 10 and 20 °C/min.

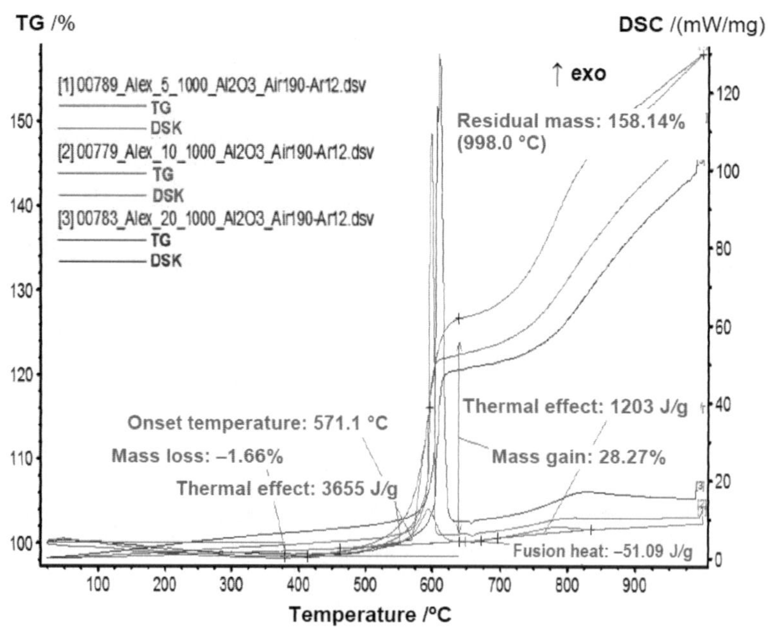

Figure 2. TG and DSC diagrams for Alex at heating rates of 5, 10 and 20 °C/min.

Changing of the heating rate in the range from 5 to 20 C/min declines the mass gain of aluminum powders: for ASD-4 – from 124.13 to 113.21%; for Alex – from 158.14 to 144.03%. This indicates that at a low rate of heating the oxidation reaction of aluminum powder has time to pass more completely.

Analysis of the results obtained with DSC analysis showed that the energy of heat release during oxidation of ASD-4 aluminum powder decreases from 326.5 J/g to 248.8 J/g with increasing heating rate, becoming more convex but narrow in the shape. Under the above conditions, the melting heat value for ASD-4 varies from –193 J/g to –178.2 J/g. For the Alex brand powder, the energy of heat generation during oxidation changes from 3655 J/g to 3467 J/g as the heating rate increases. The heat of fusion (under identical conditions) for Alex varies from –51.09 J/g to –14.45 J/g.

The generalized results for TG and DSC analyzes are given in tables 1 and 2.

Table 1. Thermophysical characteristics of the ASD-4 oxidation process.

Heating rate °C/min	First stage of oxidation		Second stage of oxidation		Thermal effect	
	$T_{o.ox}$, °C	$T_{e.ox}$, °C	$T_{o.ox}$, °C	$T_{e.ox}$, °C	ΔH_{ox}, J/g	ΔH_{m}, J/g
5	571	635	750	up to 1000	326.5	–193.0
10	576	640	760	up to 1000	255.4	–192.7
20	576	650	830	up to 1000	248.8	–178.2

Note: $T_{o.ox}$, °C – onset oxidation temperature, $T_{e.ox}$, °C – end-point oxidation temperature, ΔH_{ox} – thermal effect of oxidation, ΔH_{m} – thermal effect of melting (fusion heat).

Table 2. Thermophysical characteristics of the Alex oxidation process.

Heating rate °C/min	First stage of oxidation		Second stage of oxidation		Thermal effect	
	$T_{o.ox}$, °C	$T_{e.ox}$, °C	$T_{o.ox}$, °C	$T_{e.ox}$, °C	ΔH_{ox}, J/g	ΔH_{m}, J/g
5	550	625	705	up to 1000	3655.0	–51.1
10	560	610	720	up to 1000	3391.0	–44.1
20	580	625	740	up to 1000	3467.0	–14.5

Translating any processes detected in thermal analysis to practical problems requires a well-justified kinetic model. It is of great practical importance to determine the parameters of the formal kinetics – the activation energy and the pre-exponential factor. These parameters are important for understanding the combustion mechanism and constructing theoretical models of combustion of high-energy materials.

The reaction rate dependence on temperature is established by the Arrhenius equation:

$$K = A \cdot \exp\left(-\frac{E_a}{RT}\right),$$

where K is the rate constant, A is a pre-exponential factor, associated with the probability of collision and E_a is the activation energy.

Based on the results of TG and DSC analyzes, the activation energy (E_a) and the pre-exponent (A) were calculated for the oxidation of aluminum powders by three formal kinetic methods according to the program Kinetics 3 of the NETZCH software (tables 3, 4).

The activation energy and the pre-exponent were obtained using model-free Friedman analysis, Ozawa-Flynn-Wall analysis and ASTM E698 analysis. To determine the activation energy according to the ASTM E698 standard, the peak of the heat flow rate were used without taking into consideration the type of reaction. Thus an average value of the activation energy was obtained. In the Friedman differential method and the Ozawa-Flynn-Wall integral method, points with the same degree of conversion of all curves (isoconversion) were used to determine E_a without taking into account type of reaction. The data of TG and DSC analyzes for the low-temperature oxidation stage at three heating rates were used to calculate the formal-kinetic parameters.

Preliminary estimates of the formal-kinetic parameters (E_a, A) for aluminum powders of different dispersion at a conversion degree of 0.2 were obtained.

The results of calculations showed the sensitivity of the program to the temperature range selection.

The obtained values of the activation energy and the pre- exponential factor are consistent with the results of [10].

Table 3. Values of activation energy and pre-exponential factor for ACD-4.

Method	Activation energy, E_a, kJ/mol	Pre-exponential factor, $\log(A/s^{-1})$
Friedman analysis	308.55 ± 38.76	18.85
Ozawa-Flynn-Wall analysis	385.65±3.43	23.36
ASTM E698	566.63 ± 6.50	34.25

Table 4. Values of activation energy and pre-exponential factor for Alex.

Method	Activation energy, E_a, kJ/mol	Pre-exponential factor, $\log(A/s^{-1})$
Friedman analysis	449.21 ± 48.23	27.51
Ozawa-Flynn-Wall analysis	358.64 ± 26.86	21.88
ASTM E698	581.96 ± 150.70	35.77

4. Conclusions

In the course of this work, thermogravimetric studies showed that aluminum nanopowder Alex provides more full reaction of oxidation than ACD-4. When heated, the mass gain of aluminum oxide for Alex in the first stage of oxidation is 1.3 times higher than that of ASD-4 under the identical conditions and the same heating rates. In the second stage, the oxidation rate the samples of ASD-4 and Alex are close. At the same time, ASD-4 has a lower degree of sample transformation even when 1000 °C. The TG and DSC analyzes presented the following results: a) the effect of the aluminum powder dispersion on the oxidation process – for Alex all the oxidation processes shift to the lower temperature region; b) the effect of the heating rate on the oxidation onset and the energy of heat release. The use of modern kinetic evaluation methods for complex reactions allows to select formal-kinetic model and to determine activation parameters. The activation energy and the pre-exponential factor were calculated. Obtained results are of interest in describing the combustion processes of aluminum powders or any condensed materials containing aluminum.

References

[1] Gorbenko T I 2007 Vliyaniye sootnosheniya komponentov na goreniye metallizirovannykh topliv pri subatmosfernykh davleniyakh *Vestnik Tomskogo gosudarstvennogo universiteta* **298** pp 125–128

[2] Arkchipov V A, Gorbenko M V and Gorbenko T I 2011 Vliyaniye kataliticheskikh dobavok na skorost' goreniya kompozitsiy na osnove beskhlornogo okislitelya v diapazone subatmosfernykh davleniy *Proc. Conf. Fundamental'nyye i prikladnyye problemy sovremennoy mekhaniki* (Tomsk: Tomsk State University Press) pp 62–63

[3] Sossi A, Duranti E, Paravan C, DeLuca L T, Vorozhtsov A B, Gromov A A, Pautova Yu I, Lerner M I and Rodkevich N G 2013 Non-isothermal oxidation of aluminum nanopowder coated by hydrocarbons and fluorohydrocarbons *Applied Surface Science* **271** pp 337-343

[4] Shevchenko V G, Eselevich D A, Ancharov A I and Tolochko B P 2014 Effect of calcium on the oxidation kinetics and phase composition of the products of interaction of aluminum-based alloy powders *Combustion, Explosion and Shock Wave* vol 50 No 5 pp 534–537

[5] Shevchenko V G, Eselevich D A, Ancharov A I and Tolochko B P 2014 Effect of barium on the oxidation kinetics of an aluminumbased alloy powder *Combustion, Explosion and Shock Waves* vol 50 No 6 pp 647–652

[6] Dreizin E L and Schoenitz M 2015 Correlating ignition mechanisms of aluminum-based reactive materials with thermoanalytical measurements *Progress in Energy and Combustion Science* **50** pp 81-105

[7] Arkhipov V A, Gorbenko T I, Gorbenko M V and Savel'eva L A 2009 Effect of ultrafine aluminum on the combustion of composite solid propellants at subatmospheric pressures *Combustion, Explosion and Shock Waves* vol 45 No 1 pp 40–47

[8] Chen L, Song W L, Wang J Lv L and Xie C S 2009 Effect of heating rates on TG-DTA results of aluminum nanopowders prepared by laser heating evaporation *Journal of Thermal Analysis and Calorimetry* **96** (1) pp 141–145

[9] Teipel U 2004 *Energetic Materials* (Germany: Wiley-VCH Weinheim)

[10] Hasani S, Panjepour M and Shamanian M 2012 Oxidation and Kinetic Analysis of Pure Aluminum Powder under Non-isothermal Condition. 1:385. doi:10.4172/scientificreports.385

IOP Conf. Series: Journal of Physics: Conf. Series **919** (2017) 012023 doi:10.1088/1742-6596/919/1/012023

Investigation of chaotic and regular dynamics of the asteroids – Venus companions

T Yu Galushina and O N Letner

Tomsk State University, Tomsk, Russia

volna@sibmail.com

Abstract. The paper deals with the investigation of regular and chaotic dynamics of near-Earth asteroids (NEAs) in the vicinity of resonance 1:1 with Venus. The degree of chaoticity of NEA orbits is determined by analyzing the behavior of parameters MEGNO and OMEGNO. It is shown that the asteroid motion in the resonance vicinity and/or multiple close approaches of NEAs with planets lead to the appearance of dynamical chaoticity.

The investigation of regular and chaotic dynamics of near-Earth asteroids (NEA) is very topical currently. One of the main reasons of particular attention to the NEA dynamics is the asteroid hazard problem. The mean motion resonances with major planets are of great importance in NEA motion because stable resonances can be a protection mechanism from asteroid close encounters. However, the motion in the vicinity of the resonances can lead to significant changes of asteroid orbital parameters and to possible close approaches with the planets, particularly with the Earth. The investigation of NEA chaotic dynamics allows one to determine a time interval of asteroid motion predictability and the time after that the motion may be considered as unpredictable.

This paper presents some results of investigating regular and chaotic dynamics of the asteroids being Venus companions which are revealed at the beginning of 2015. There were five such objects at the time of the revealing – 322756 2001 CK32, 2002 LT24, 2002 VE68, 2007 AG, 2013 ND15 [1].

For all the investigated NEAs the orbital evolution has been simulated over a time interval about 3000 years. The behavior of resonant characteristics and chaoticity parameters has been studied over the same span. The investigation has been carried out by numerical integration of motion differential equations by Everhart method [2]. The initial data have been taken from Bowell catalog dated February of 2015 [3]. The force model includes the gravitational influence of the planets, Pluto, the Moon, Ceres, Pallas and Vesta. For all the computations has been used a specially developed software [4] which enables one to perform high-precision prediction of asteroid motion.

We used the following resonant characteristics: a critical (resonant) argument β which determines the longitude of the connection for an asteroid and Venus [5]:

$$\beta = \lambda_1 - \lambda_2, \tag{1}$$

and its time derivative α (called resonant "band") [6]:

$$\alpha \approx n_1 - n_2, \tag{2}$$

where n_1, n_2 are the mean motions, λ_1, λ_2 are the mean longitudes of the asteroid and Venus respectively. At $\alpha = 0$ the asteroid is in the exact resonance with the planet due to the mean motion commensurability 1:1. The asteroid moves in the vicinity of the resonance if α and β librate around the value of the exact commensurability so that $|\beta - \bar{\beta}| \leq 180°$ and $|\alpha| \leq \alpha_{max}$ ($\bar{\beta}$ is the libration center for the critical argument). The value α_{max} characterizes boundaries of the resonant motion and is determined by the maximum amplitude of the critical argument librations.

We used the indicators MEGNO (Mean Exponential Growth factor of Nearby Orbits) [7] and OMEGNO (Orthogonal MEGNO) [8] to study regularity or chaoticity of orbital motion in the vicinities of resonances. Analysis of these characteristics allows separating regular and chaotic regimes of motion surely.

For example, the averaged parameter MEGNO $\bar{Y}(t)$ has the reference value 2 relative to which the chaoticity and regularity are determined. For chaotic orbits with exponential divergence of close trajectories the averaged parameter MEGNO is greater than 2 and it grows almost linearly. For stable quasi-periodical (regular) orbits with linear divergence of close trajectories $\bar{Y}(t)$ always tends to 2 and for stable orbits described by equations for a harmonic oscillator $\bar{Y}(t)=0$ [9].

The orthogonal parameter MEGNO, i.e. the parameter OMEGNO $Y_\perp(t)$, allows not only distinguishing chaotic and regular orbits but also revealing and separating periodic trajectories in the domains of regular motion of continuous dynamic systems [10]. This is an advantage of the given parameter.

The parameter OMEGNO has two reference values $Y_\perp(t)=2$ and $Y_\perp(t)=0$. When $Y_\perp(t)>2$ and the parameter OMEGNO grows almost linearly, a motion is considered as chaotic. For quasi-periodic (regular) orbits $Y_\perp(t)\to 2$ and the deviations of $Y_\perp(t)$ from 2 allows determining whether the trajectory under investigation is close to a stable or unstable periodic orbit. That is to say, if $Y_\perp(t)\lesssim 2$ then the trajectory is close to a stable periodic orbit and if $Y_\perp(t)\gtrsim 2$ then the trajectory is close to a unstable periodic orbit. At last, for any periodic orbit $Y_\perp(t)=0$.

For each asteroid, the graphs of evolution of the orbital elements (semi-major axis *a*, eccentricity *e*, and inclination *i*), the resonant argument β (1), the resonant band α (2), the close encounters with planets, and also the graphs of evolution of the chaoticity parameters MEGNO and OMEGNO have been constructed and analysed.

The investigation results showed that all the five NEAs move in the vicinity of resonance 1:1 with Venus and have non-regular librations of the resonant band around the value of the exact commensurability of the mean motions of the asteroid and the planet. Only asteroid 322756 2001 CK32 moving on one side of resonance in past falls into the resonance in 2000 and regularly passes through the exact commensurability in future. This object has a quasi-periodic (regular) character of motion over the entire considerable time interval, as is followed from the evolution of the MEGNO and OMEGNO parameters. At the same time four from the five NEAs begin to show chaoticity features. The transition between resonant and nonresonant motions and close and/or multiple approaches of these asteroids with planets lead to the appearance of motion chaoticity.

As an example figure 1 and figure 2 present orbital evolution of asteroids 322756 2001 CK32 [11] and 2002 LT24 [12] respectively. Every figure shows evolution of the resonance band α (*a*), evolution of the critical argument β (*b*), approaches with Venus (*c*), the Earth (*d*), Mercury (*e*), evolution of the OMEGNO (gray line) and averaged MEGNO (black line) parameters (*f*), evolution of the semi-major axis *a* (*g*), the eccentricity *e* (*h*), and the inclination *i* (*i*).

As is seen from Figure 1, the critical argument of asteroid 322756 2001 CK32 moving along one side of the commensurability circulates but does not librate. From 1250 the parameter MEGNO oscillates near 2, i.e. from that time the motion of 322756 2001 CK32 is regular. Moreover, evolution of the parameter OMEGNO shows that the asteroid orbit is close to a stable periodic orbit, $Y_\perp(t)\to 0$. The transition to the resonant motion leads to a jump of both chaoticity parameters but, nevertheless, the object motion remains quasi-periodic. At the same time $Y_\perp(t)\to 2$ from above, i.e. the orbit is close to unstable periodic.

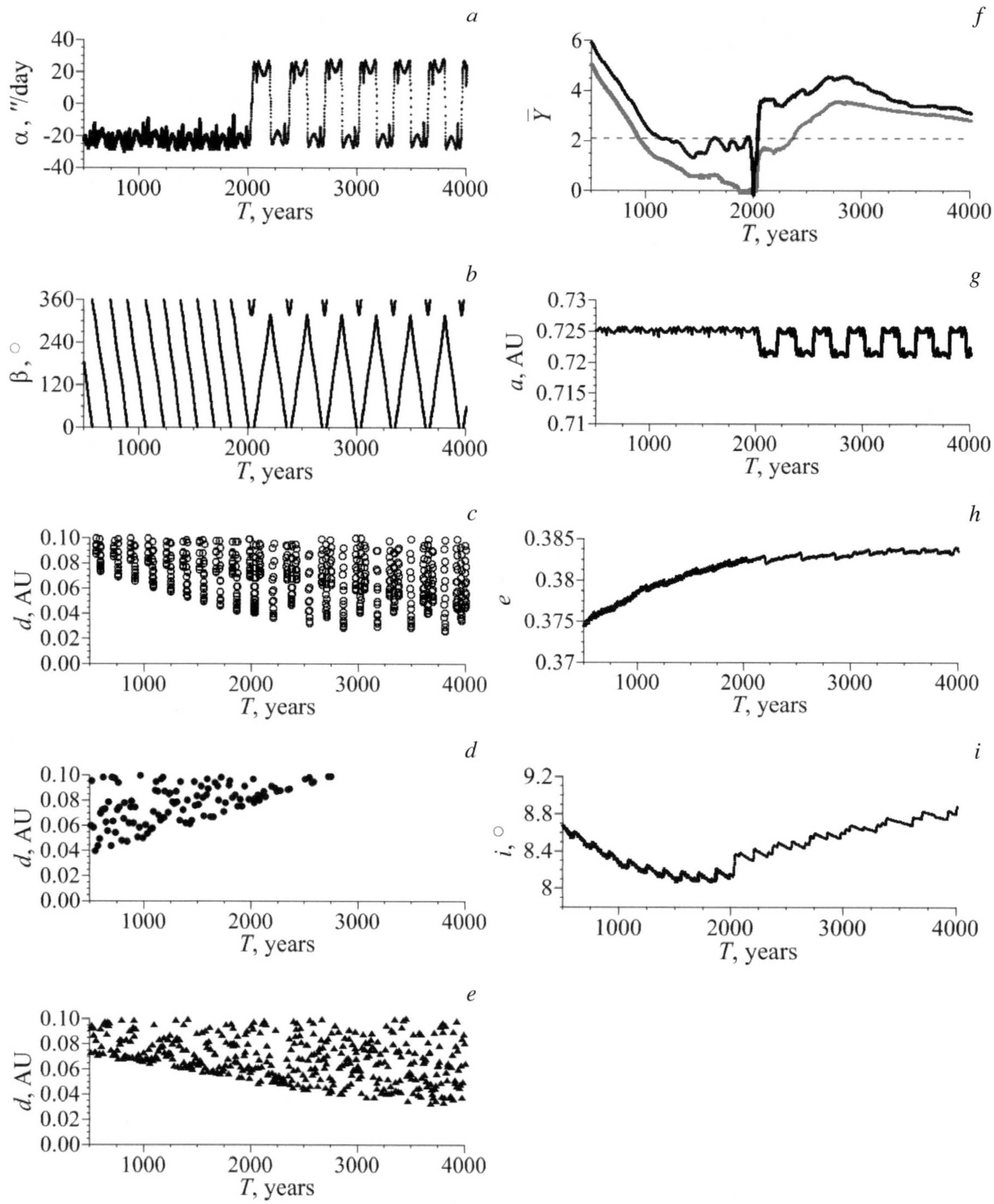

Figure 1. Asteroid 322756 2001 CK32: evolution of the resonance band α (*a*), evolution of the critical argument β (*b*), approaches with Venus (*c*), the Earth (*d*), Mercury (*e*), evolution of OMEGNO (gray line) and averaged MEGNO (black line) parameters (*f*), evolution of the semi-major axis a (*g*), the eccentricity (*h*), and the inclination of the asteroid orbit i (*i*)

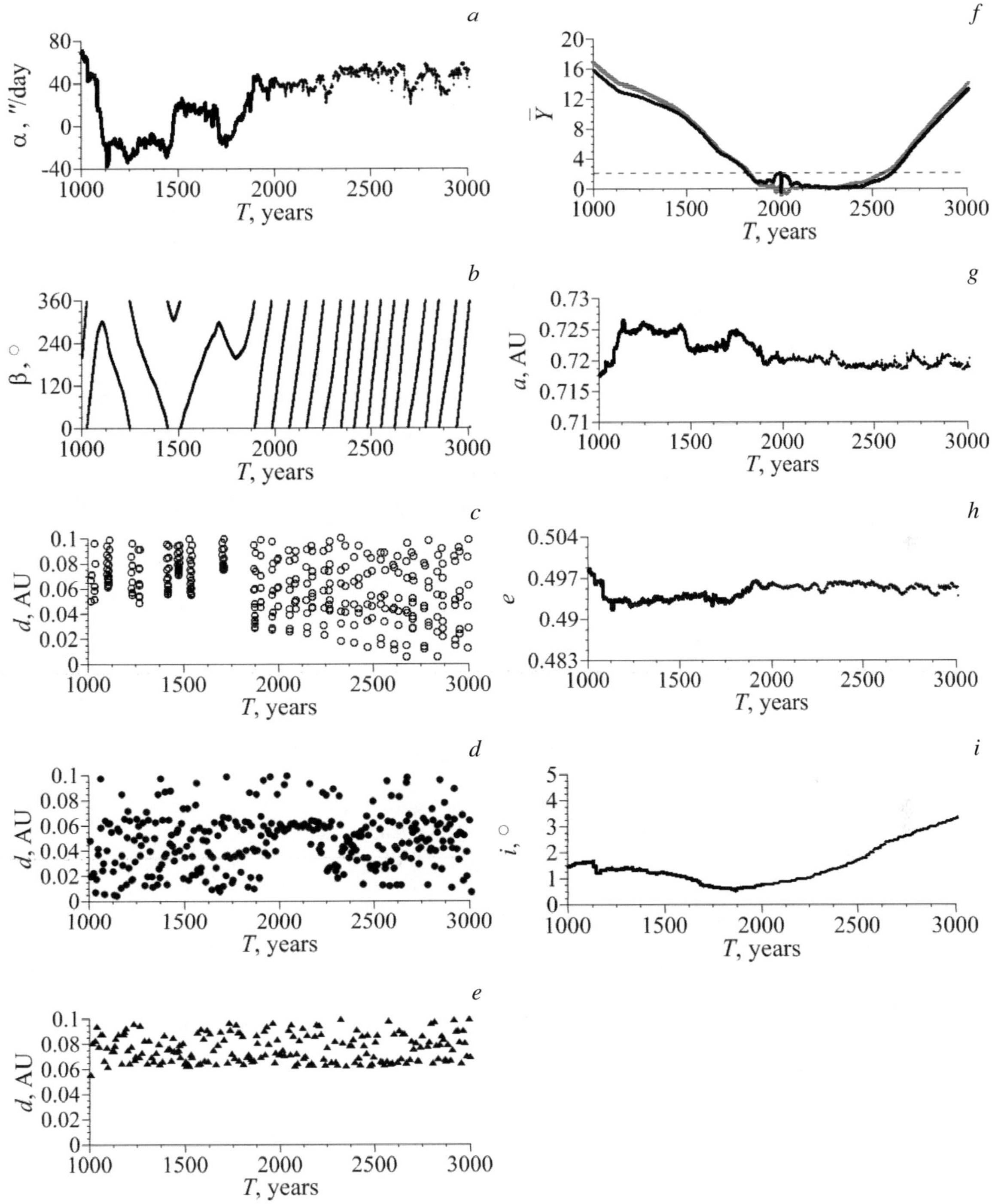

Figure 2. The same as in Fig. 1 but for asteroid 2002 LT24

Figure 2 shows that the asteroid 2002 LT24 moves on one side of the exact commensurability most of the time. The resonant band changes its sign only several times and the critical argument circulates almost always. The parameters MEGNO and OMEGNO are close to zero from 1800 to 2500, i.e. its orbit is stable and periodic during about 700 years.

From the graphs of close encounters of 2002 LT24 with the Earth and Venus we can see that from 2500 the asteroid has a series of approaches with the planets, which leads to the development of chaoticity. During the orbit simulation towards the past the chaoticity reveals itself at the transition from the libration to the circulation of the critical argument. Also we can see that the orbital elements of both the asteroids change insignificantly over the considered time interval.

For asteroids 2002 VE68 [13] and 2013 ND15 [14] we carried out the orbit fitting by the least square method using all available optical observations and investigated their probabilistic orbital evolution [1]. The analysis of behavior of the chaoticity parameters shows that at $Y_{\perp}(t)>2$, $\bar{Y}(t)>2$, and their further growth the probabilistic domains increase significantly, which allows making the conclusion about exponential extension of the domains.

Thus the regular and chaotic dynamics of asteroids 322756 2001 CK32, 2002 LT24, 2002 VE68, 2007 AG, 2013 ND15, Venus companions, has been investigated. All these asteroids move in the vicinity of resonance 1:1 with Venus and have irregular librations of the resonance band near the value of the exact commensurability of the mean motions of the asteroids and the planet. The results of the evolution of the parameters MEGNO and OMEGNO are in a good agreement with each other but the parameter OMEGNO behaves more smoothly and makes it possible to reliably identify a stable periodic motion. The paper shows that the transition to a chaotic motion may occur at close and multiple encounters with planets and also at the entrance in a resonant domain and at the exit from it.

This work was supported by the D.I. Mendeleev Scientific Foundation of Tomsk State University (project 8.1.28.2017).

References

[1] Galushina T, Skripnichenko P, and Titarenko E 2017 *Russian Physics Journal.* **59 9** 1401–09
[2] Everhart E 1974 *Celest.Mech.* **10** 35 – 55.
[3] Bowell E, Muinonen K, and Wasserman L 1994 *In Asteroids, Comets, Meteors, Kluwer, Dordrecht, Netherlands*. 477–81
[4] Bykova L, Galushina T, and Baturin A 2012 *Izv. Vyssh. Uchebn. Zaved. Fiz.* **55 10/2** 135–42 (In Russian)
[5] Murray C and Dermott S 1999 *Cambridge: Cambridge University Press*. 592 p.
[6] Grebenikov E and Ryabov Yu 1978 *M.: Nauka*. 128 p. (In Russian)
[7] Cincotta P and Simo C 2000 *Astron. Astrophys. Sup.* **147** 205–28.
[8] Shefer V and Koksin A 2013 *Izv. Vyssh. Uchebn. Zaved. Fiz.* **56 6/3** 256–58. (In Russian)
[9] Cincotta P and Girdano C, and Simo C 2003 *Physica D*. 151–78
[10] Shefer V and Koksin A 2016 *Russian Physics Journal.* **59 1** 71–75
[11] Brasser R, Innanen K, Connors M, Veillet C, Wiegert P, Mikkola S, and Chodas P 2004 *Icarus.* **171 1**. 102–09.
[12] Nugent C, Mainzer A, Masiero J, Grav T, and Bauer J 2012 *The Astronomical Journal* **144 3**. article id. 75, 6 pp.
[13] de la Fuente Marcos C and de la Fuente Marcos R 2012 *Monthly Notices of the Royal Astronomical Society.* **427 1** 728–39.
[14] de la Fuente Marcos C and de la Fuente Marcos R 2014 *Monthly Notices of the Royal Astronomical Society.* **439 3** 2970–77.

IOP Conf. Series: Journal of Physics: Conf. Series **919** (2017) 012024 doi:10.1088/1742-6596/919/1/012024

Problem of development high-precision local model of quasigeoid

K V Shapovalova

Tomsk State University, 36 Lenin Street, Tomsk, 634050 Russian Federation
E-mail: shapovalova.kseniya@stud.tsu.ru

Abstract. The description and the topicality of the geodetic heights transformation to the quasigeoid reference surface were shown. The problem-solving techniques, the level of the topic scientific development and perspectives for further research of the local Height Reference Surface modeling are under consideration.

Introduction

At the present day, the main and the most available method of high-precision determination of heights (elevations) in their physical meaning is geometric leveling. This method assumes direct observation, as elevations are determined relatively to the surface normal to vertical. With the active implement of satellite co-ordinate technologies the determination of heights based on observations of global navigation satellite systems becomes an alternative to the geometric leveling. The main disadvantages of the method are indirect determination of values and relative altitudes referred to reference ellipsoid, which called geodetic heights. In the Russian Federation, the quasi-geoid surface is assumed as the reference surface and heights are referred to Baltic normal height system. This fact excludes the possibility of implementation the satellite determination technology without strict relations to the reference surfaces. At the planetary scale, the connection of the systems is realized through the geoid models, which consist of a set of elevations geoid heights above the ellipsoid on a regular grid.

Therefore the presence of quasi-geoid model assumes knowing the differences between the geodetic and normal heights in any point of the planet or, at least, in nodes equiangular grid covering the Earth. At the present day, a maximum error of the quasi-geoid surface determination is in the order of 1m, which is obviously insufficient for the high-precision leveling. It was found that in order to calculate the heights of global quasi-geoid with sufficient accuracy it is required to determine gravity anomalies at a level unobtainable for advanced gravimetric equipment [3, 6]. Therefore, the development of more accurate local quasi-geoid models allowing the satellite co-ordinate determination technologies fully implemented in a limited area remains urgent. However, current regulations and standards is out of date and leaves out capabilities of modern measuring and computing equipment. That is why it does not imply the accomplishment of similar works, which seriously complicates practical implementation of the local models.

Within the problem under consideration the reduction of geodetic heights to the surface of quasi-geoid results in the constructing of more accurate model which makes possible using of satellite data for high-precision leveling.

A brief review of modern geopotential models

IOP Conf. Series: Journal of Physics: Conf. Series **919** (2017) 012024 doi:10.1088/1742-6596/919/1/012024

Until recently, the most accurate global model of Earth's gravity field was EGM96. The model is based on the set of expansion coefficients of Earth's gravity field according to spherical harmonics to 360 degrees and includes anomalies over a regular 30-minutes grid. A serious step increasing the accuracy of the model was GRACE mission satellite launch in 2002 year aimed at the Earth's gravity field and time variations research. The GRACE satellites, which operation was planned in 2016, now represent a couple of satellites that follow one after another at the distance about 220 km along a single polar orbit. The object of study is the distance between them, which measured with the accuracy of 10 microns. Proper motion and orientation of satellites are recorded with the help of GPS receivers, accelerometers and star sensors. In addition, the satellites are equipped with corner reflectors for use in satellite laser altimetry. By changing the distance between satellites, it can be inferred about the changes in the gravitational field of the Earth. One of the basic tasks of this project is the representation of both permanent and time-varying components of the Earth's gravitational field. After the launch of the GRACE mission, many scientific groups began to process the obtained data. The last available solution dates September 2016.

A new stage in the development of Earth's gravity models came with the launch of research satellite GOCE of the European Space Agency in 2009. Its distinctive feature was the record low orbit located in upper atmosphere at a height of 254 km. This feature allowed to obtain high spatial resolution; however, it led to the need to provide the satellite with a constantly working xenon ion engine to compensate for the effect of the atmospherical braking. The main useful load of satellite undertook 6 high-precision accelerometers. After working out the main observation program, the satellite had a significant fuel reserve, and the European Space Agency decided to continue the mission with a gradual orbit lowering by another 30 km. The orbit lowering further increased the spatial resolution and quality of the data obtained. Many scientists even called this the second mission. A 35 percent improvement in data quality was expected. Even in the first months of the mission after the publications of the data, it became clear that the accuracy of the results exceed the accuracy of all early observations including GRACE. More than 4 years later after fuel depletion the satellite was ocean dumping. The received data are currently processed by the world scientific community.

Today there are about a hundred geoid models, some of them are under development. The most popular and useful models are considered below.

- EIGEN-6C4 is a combined gravity field model calculated to a degree and order of 2190 using GRACE, Lageos, GOCE data and ground gravity measurements.
- TIM R5 is a gravity field model, calculated to a degree and order of 280 using only GOCE data.
- EGM2008 is a model of spherical harmonics of Earth's gravity field, calculated to a degree and order of 2159 using different data.

In the current study, we consider the model EGM2008 in details. At the present, a group of scientists from the United States, headed by N.K. Pavlis, is engaged in development and evaluation of Earth's geopotential model EGM2008. The produced model implementing different data is actively used in geodesy [8]. EGM2008 is a model of the spherical harmonics of the Earth's gravitational field, calculated to a degree and order of 2159 and containing additional coefficients to the degree of 2190 and order of 2159. This model replaced the EGM96. The EGM2008 includes gravity anomalies over a global equiangular 5-minute grid. This model is built based on surface data, takes advantages of satellite solutions for GRACE and contains improved altimetry-derived gravity anomalies calculated using PGM2007B (version PGM2007A) and dynamic ocean topography (DOT) models. The development of EGM2008 faced with some difficulties related to data insufficiency. For example, the altimetry-derived gravity anomalies in free air are combined with the corresponding values over the earth's surface and supplemented with «completed» values for regions without any gravitational observations to form a complete global equiangular 5-minute grid of gravity anomalies in free air.

Thus, currently EGM2008 is one of the best models but still its accuracy is not enough for the high-precision levelling.

Many specialists work on the EGM2008 model local elaboration. A series of the most significant studies is given below.

V.Z. Ostroumov worked on development and evolution of methodology for determining the heights of tide gauge stations based on satellite observations using local and global quasi-geoid models from 2004 to 2013. In his articles are given the comparative characteristics of some global geoids and methods for their refinement on a local site. One of the articles is entirely devoted to the problem of increasing the accuracy of determining the altitude component of the coordinates obtained from GNSS data [1].

A local geiod model for the Novosibirskaya region was calculated in Siberian State University of Geosystems and Technologies [2]. The project was implemented based on a network consisting of 31 continuously operating reference stations evenly covering entire territory. The model is also obtained by refining the global model EGM2008. Geodetic and normal heights were taken from the materials by adjustment this network, but the normal heights of the base stations were not adjusted together, because it was obtained from isolated courses of geometric 1-4-order leveling. It is noteworthy that the project used ready-made commercial software.

Well-known A.N. Majorov model proved the possibility of determining normal heights with the accuracy of geometric levelling of III – IV order according to observations of satellite systems [5].

Vu Hong Cuong developed 12 variants of quasi-geoid models for Vietnam by combining the coefficients of high and low harmonics from the global geoid models GOCE-TIM4, EGM2008 and EIGEN-6C in Moscow State University of Geodesy and Cartography [4].

Algorithms of calculating height anomalies by combining gravity measurements, global gravity models, normal and geodetic heights were developed by D.A. Shogonbekova in 2015 in Kazakh National Research Technical University named after K.I. Sitpaev. Based on the results of works, the geoid model was developed for the territory of Kazakhstan. The model is based on the global model EIGEN-6C4 with a standard deviation of 0.037 m. [7].

There are other models developed for different local regions of the Earth's surface but the determination of normal heights using these models also does not to exceed the accuracy of geometric levelling of III – IV order.

Conclusion

All works reviewed above were carried out with the use of data taken from the list of coordinates and heights of bench marks adjusted many years ago but with up-to-date satellite observations for these points. Due to permanent geodynamical and technogenic effects on bench marks their heights have to be re-determinate. Consequently, an insufficiency of effective data for geometric levelling is a week point in these works. In addition to out-of-dateness of reviewed data, it should be considered that at the moment there is a more advanced equipment that allows making measurements in a semi-automatic mode and reducing operator and other random errors manifold. As geometric levelling data are the standard for local elaboration of global geoid models, the key role in this process should be taken by procedure of levelling data updating.

Nevertheless a lot of digital geodetic levels imported to Russia, at the moment there is no regulations and standards for working with this device. However, there are works devoted to the influence of various sources of errors on digital geodetic levels and bar-code leveling rods. It can be noted that such geodetic levels are more sensitive to external factors and should be used with caution. In this connection, it is preferably to take into account the data of experiments on the behavior of digital geodetic levels in different environments. U.E. Fedoseev analyzed the existing regulations and standards in the area of geometric leveling, and also tried to interpret it to work with digital geodetic levels. In his paper [8], all the main problems that arise when working with this electronic device are reviewed in detail. A significant problem is also the metrological certification of digital geodetic levels. In connection with the absence of a bar-code leveling rod of the metric scale, control of meter and decimeter intervals of the leveling rod, assumed by the current standard for leveling, is a very sophisticated problem, and according to the article by N.K. Goligin, etc. [9] generally loses its meaning, since the bar-code is read from more than 300 millimeter leveling rod section. The absence of a metric scale does not allow in the field environment to reject measurements containing gross errors, and the current regulations and standards does not contain suitable control procedures.

To solve our problem it is necessary to build a classical high-precision leveling network around the perimeter of the city, including the laying of missing geodetic marks. This type of leveling network is a universally recognized high-precision construction and will serve as a standard. The done extensive redundant observations of global navigation satellite systems on the points of this reference network will allow us to collect a sufficient amount of statistical information to determine the limits to applicability of the technologies themselves. This approach will allow to determine the minimum amount of measurements necessary to solve a particular problem. The collected data will be used in the preparation of regulations and standards for the work. When observational data processing, it is necessary to analyze several progressive mathematical algorithms taking into account the specific nature of new types of measuring instruments.

The solution of this problem is a priority, as it leads to a significant costs saving when geodetic support of civil construction, reduces the influence of the human factor and increases the automation of production processes.

The possibility of using satellite observations in high-precision levelling gives the new opportunities for Russian and foreign global navigation satellite systems.

References

[1] Ostroumov V Z 2013 Determination of Normal Heights of the Tide Gauge Stations' Benchmarks by the GNSS Data and the Quasigeoid Model in the Azov and Black Sea Region *Geoprophy* **3** 20-23

[2] Obidenko V I, Opritova O A and Reshetov A P 2016 Working out of a technique of reception of normal heights in territory of the Novosibirsk region with use of earth gravitanional model EGM2008 *Vestnik SSUGT* **1 (33)** 14-25

[3] Kanushin V F, Karpik A P, Goldobin D N, Ganagina I G, Gienko E G and Kosarev N S 2015 The definition of gravity potential and heights differences in geodesy by gravimetric and satellite *Vestnik SSUGT* **3 (33)** 53-69

[4] Vu C H 2013 *Development of methodology for constructing a quasi-geoid from satellite measurements in Vietnam Extended abstract of PhD dissertation* Moscow State University of Geodesy and Cartography Moscow

[5] Majorov A N 2008 *Development of technology and creation of a quasi-geoid model using satellite data PhD dissertation* Moscow State University of Geodesy and Cartography Moscow

[6] Dementjev U V 2012 *Evolution of the theory and development of technology for determining the anomaly of gravity in complete topographic reduction doctoral dissertation* Siberian State Geodetic Academy Novosibirsk

[7] Shoganbekova D A 2015 *Development of algorithms for calculating heights anomalies and modeling gravimetric geoid of the Republic of Kazakhstan PhD dissertation* Kazakh National Research Technical University named after K.I. Sitpaev Almaty

[8] Fedoseev U E 2010 The application of digital levels for high-precision work: theory, practice, problems *Automated technologies of survey and design* **2 (37)**.

[9] Goligin N H, Fedoseev U E and Cherepanov P A 2013 Perspectives for using measuring systems "digital level + bar-code leveling rod" *News of Higher Educational Institutions «Geodesy and air-photography »* **6** 13-16.

[10] Pavlis N K, Holmes S A, Kenyon S C and Factor J K 2012 The development and evaluation of the Earth Gravitational Model 2008 (EGM2008) *Journal of geophysical research* **117** 1-38.

[11] Fotopoulos G 2005 Calibration of geoid error models via a combined adjustment of ellipsoidal, orthometric and gravimetric geoid height data *Journal of Geodesy* **79 (1-3)** 111-123

[12] Shapovalova K V 2014 *Int. youth scientific Conf. «Current issues of continuum mechanics and celestial mechanics – 2014» (Tomsk)* vol 1 (Tomsk: National Research Tomsk State University) p 135

IOP Conf. Series: Journal of Physics: Conf. Series **919** (2017) 012025 doi:10.1088/1742-6596/919/1/012025

Evaluation of influence of different perturbing accelerations on accuracy of probabilistic model of asteroid 2011 MD motion

O M Syusina, G E Sambarov

Tomsk State University, 634050 Tomsk, Russian Federation

E-mail: kleo77@sibmail.com

Abstract. This paper is devoted to investigation of the influence of various perturbing factors on accuracy of probabilistic model of asteroid 2011 MD motion. The degree of influence of each perturbing acceleration has been estimated using indicator of the force model. This indicator has been determined on the basis of the algorithm in which the errors of force models are tightly connected with sizes of confidence regions and displacements of the least-square estimations defined for different models of motion of asteroids. For object 2011 MD the influence of the Sun oblateness and relativistic effects caused by the Sun has been of less importance then the influence of gravitational perturbations from the Moon and the Earth oblateness. The investigation probabilistic orbital evolution on the basis of different force models has shown that the neglect of perturbing accelerations whose indicator of the force model is greater than the threshold value can lead to divergence of orbits.

1. Introduction

Investigation of orbital evolution of the small Solar System bodies, as well as the task of probability evaluation of the object collision with the Earth, require making up a perturbing accelerations model influencing the object under research. The force model used in differential equations of asteroids motion can contain different set of perturbing accelerations depending on an object orbit type and its possible encounters with other planets. The application of a less detailed model than is needed, leads to the non-occurrence of the object under study into the confidence region of the object motion. This is highly unacceptable when dealing with task of probability evaluation of asteroid collision with the Earth. In the articles [1-3] we have proposed the way of evaluation of accuracy of asteroid motion model based on calculation indicator of displacement relative to each other confidence regions constructed for different object motion models. In the present study this way has been applied to the force model for the asteroid 2011 MD which is approaching the Earth.

2. Accuracy indicator of the force model

When investigating the probabilistic evolution of the small Solar System bodies, one can evaluate the influence of perturbing accelerations included in the force model for this object, with the help of the following indicator (1).

$$\varepsilon = \frac{\left|\widehat{\mathbf{q}} - \widehat{\mathbf{q}}^*\right|}{\left|\overline{\mathbf{q}} - \widehat{\mathbf{q}}^*\right|} = \frac{1}{\overline{l}}, \qquad (1)$$

where $\widehat{\mathbf{q}}$ is the vector of asteroid orbital parameters $\mathbf{q} = (q_1, q_2, ..., q_m)$, obtained by the least square method for model of perturbing accelerations F; $\widehat{\mathbf{q}}^*$ is the same, but for model F^*; $\overline{\mathbf{q}}$ is a point lying in the parametric space along the direction $(\widehat{\mathbf{q}} - \widehat{\mathbf{q}}^*)$ on the boundary surface of the asteroid's motion confidence region $\Phi(\mathbf{q}) = \Phi(\widehat{\mathbf{q}}^*)(1 + k^2/(n - m))$; $\Phi(\widehat{\mathbf{q}}^*)$ is mean square error for the full model; the value k is estimated using the Fisher distribution statistics; n is number of object observations.

The given indicator allows to take into consideration the size of confidence regions and shifts of parameters estimation of the object orbits, determined for a complete (F) and incomplete (F*) disturbing force models. This indicator can be applied not only for confidence regions at the initial time, but also for their mapping in time. The value $\varepsilon > 1$ obviously denotes significant shifts of parameters estimation $\widehat{\mathbf{q}}$ and $\widehat{\mathbf{q}}^*$ relatively to each other. Comparison of the received value ε with some threshold value $\overline{\varepsilon}$ allows to evaluate the degree of perturbing accelerations influence. According to our numerical evaluation, the value $\overline{\varepsilon}$ should be chosen within the interval $0.01 < \overline{\varepsilon} < 0.1$ [4-5].

Other methods for the investigation of the perturbations structure are given in the works [6-7].

3. Numerical experiment

A near-Earth asteroid 2011 MD has been chosen as an object of the study. This object has 1555 observations in the database of IAU Minor Planet Center (MPC) performed within the interval from June, 21st to September, 3rd, 2011. On June, 27th, 2011 the asteroid approached the Earth surface relatively close – at the distance equal to 18000 km from the centre of the Earth. The complete force model for the given object contained gravitational perturbations from all major planets, Pluto, the Moon; the Earth, the Sun and Jupiter oblateness; relativistic effects caused by the Sun. With the help of the indicator ε (1), the influence of each perturbing acceleration included in the force model has been evaluated. The value $\overline{\varepsilon} = 0.01$ has been taken as a threshold one. Table 1 contains the ε indicator values, evaluating the influence gravitational perturbations from the Moon; relativistic effects caused by the Sun; the Earth and the Sun oblateness. Indicators have been obtained within the object's interval of observations.

Table 1. The accuracy indicator ε of the force model.

Perturbation	ε
The Moon	13.32
The Earth oblateness	0.461
The Sun oblateness	$1\cdot10^{-6}$
The relativistic effects of the Sun	$1\cdot10^{-5}$

The table shows that the indicator ε value evaluating the influence gravitational perturbations from the Moon and the Earth oblateness exceeds the $\overline{\varepsilon}$ threshold value within the interval of the observations. This means that the given perturbing accelerations should be taken into consideration when researching motion of this object. The indicator value ε evaluating the influence of the Sun oblateness and relativistic effects caused by the Sun is considerably below the threshold value.

Figure 1 and 2 gives a clear presentation of confidence regions shift when using different force models. Black color in both figures denotes an initial confidence region, which is constructed by a

IOP Conf. Series: Journal of Physics: Conf. Series **919** (2017) 012025 doi:10.1088/1742-6596/919/1/012025

boundary surface [5] with the use of a complete model of disturbing forces. Gray color denotes an initial confidence region constructed with the use of incomplete model regardless of the Earth oblateness (figure 1) and regardless of the Sun oblateness (figure 2). Figures 1 and 2 are constructed with respect to the solution obtained on the basis of the complete motion model.

The figure depicts that in the case when we do not consider perturbing accelerations from the Earth oblateness, the regions displace relatively one another, and in the case when we do not consider perturbing accelerations from the Sun oblateness – the regions superimpose on one another. These results are in good agreement with the indicator value ε given in the table 1.

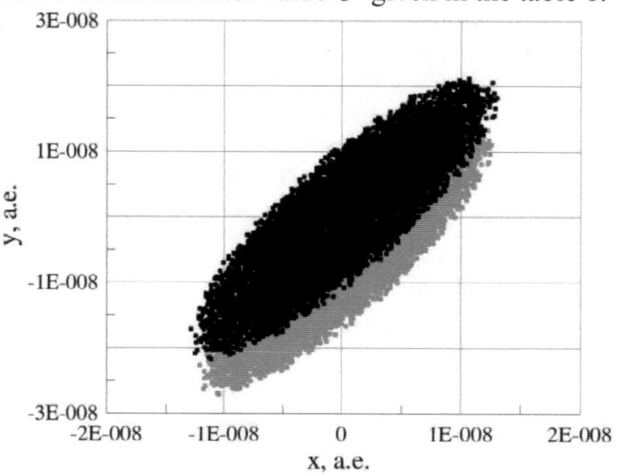

Figure 1. Initial confidence regions for complete model and incomplete model regardless of the Earth oblateness

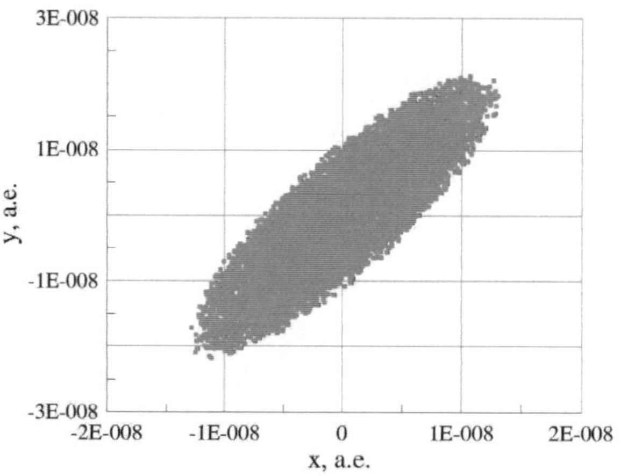

Figure 2. Initial confidence regions for complete model and incomplete model regardless of the Sun oblateness

The further study of probabilistic orbital evolution was conducted by numerical integration of differential equations of motion for the next 100 years. That has been done for nominal orbits received on the basis of different force models. Below, figure 3 depicts evolution of the semi-major axis a, the eccentricity e, the inclination of the orbital plane to equator i, the longitude of the ascending node Ω, the argument of perihelion ω. Figure 4 shows the closest approaches of the object to the Earth within the analyzed time interval. Here, gray color marks the results obtained on the basis of motion model regardless of the Earth oblateness, and black color marks the results obtained on the basis of a complete force model.

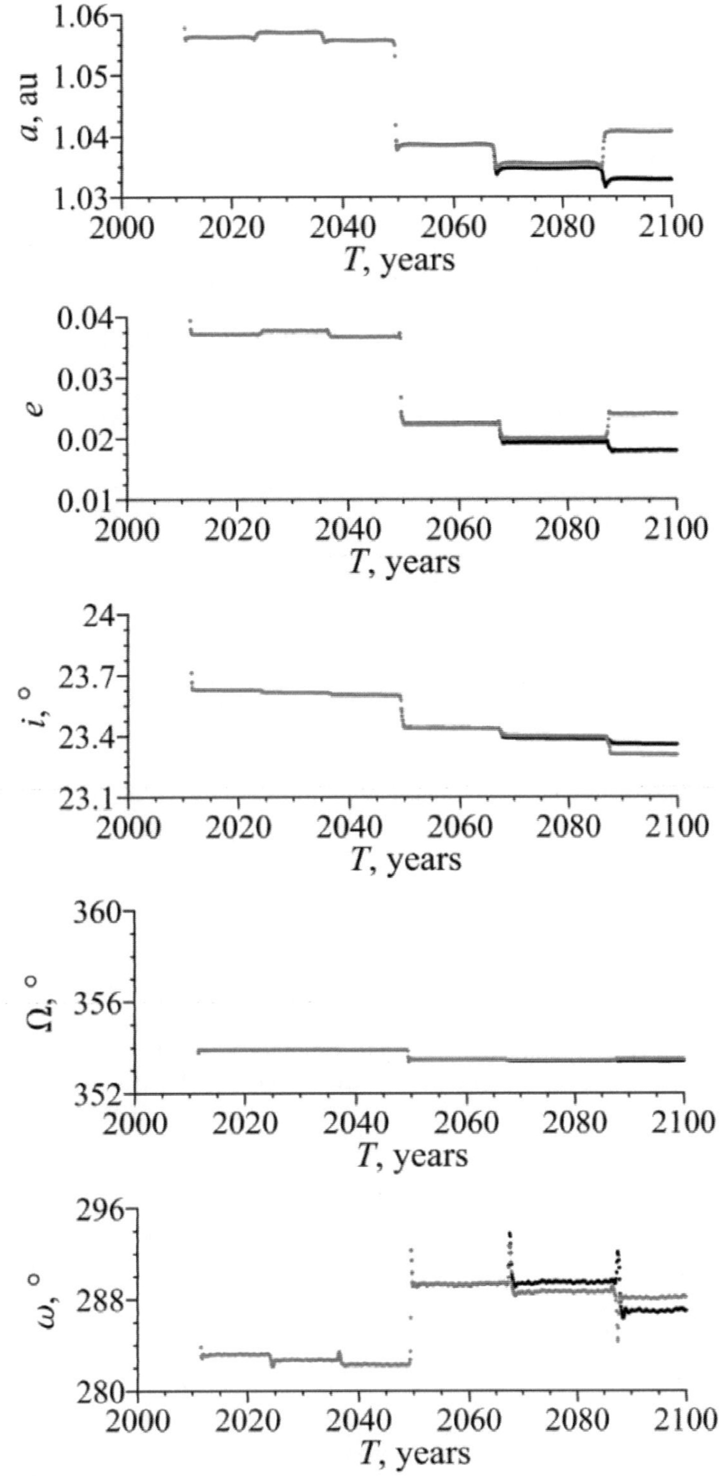

Figure 3. The results of the probabilistic orbital evolution investigation for asteroid 2011 MD

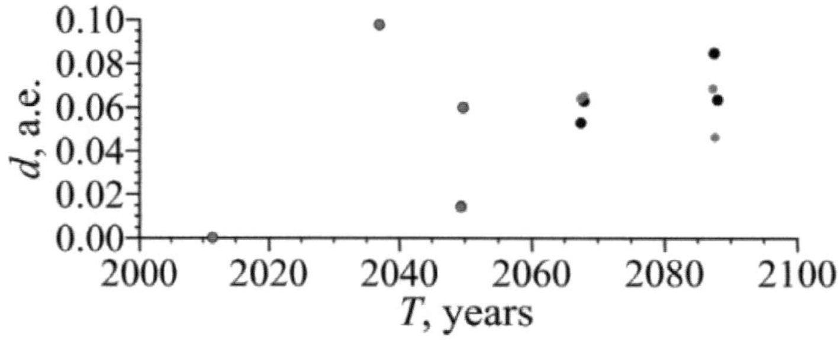

Figure 4. The closest approaches of the object to the Earth

As we can see from the figure, after the year 2064 two orbits have significant spatial discrepancies. Moments of encounters to the Earth and distances between the object and the planet after this year also vary (figure 4).

4. Conclusion

The indicator ε allows to evaluate the influence of perturbing accelerations on accuracy of probabilistic model construction of asteroids motion. This method can be applied both in making up initial confidence regions and in displaying these regions in time. As a whole, the picture of influence of the perturbing factors on accuracy of confidence regions construction can be complicated and may depend on close approaching of asteroids to the Earth or the other planets.

Acknowledgements

This work was supported by grant of the Russian Foundation for Basic Research (project № 16-32-00191 mol_a).

References

[1] Sambarov G E, Chernitsov A M and Tamarov V A 2012 Estimating the effect of Earth and Sun oblateness on the accuracy of probabilistic motion model of asteroids, observed in one opposition *Izv. Vuzov. Fiz. 57 (10/2)* (Tomsk: Tomsk State University) pp **33–42** (in Russian)

[2] Syusina O M, Chernitsov A M, Tamarov V A and Sambarov G E 2013 Method of estimation of force model influence on systematic error of determination regions of probable asteroids motions *Izv. Vuzov. Fiz. 56 (6/3)* (Tomsk: Tomsk State University) pp **235–237** (in Russian).

[3] Sambarov G E, Syusina O M 2016. Investigation of the probabilistic model of motion of near-Earth asteroids *Collection of works IX all-Russian scientific. conf. «Fundamental and applied problems of modern mechanics»* (Tomsk: Tomsk State University) pp **351–353** (in Russian)

[4] Syusina O M, Sambarov G E, Chernitsov A M and Tamarov V A 2012 Investigation of influence of the asteroid motion model errors on precision of the building of initial confidence region *Izv. Vuzov. Fiz. 55 (10/2)* (Tomsk: Tomsk State University) pp **50–59** (in Russian).

[5] Syusina O M, Chernitsov A M and Tamarov V A 2012 Construction of confidence regions in problem on probabilistic study into motion of minor bodies of the solar system *Sol. Syst. Res. 46 (3)* pp **195–207**.

[6] Galushina, T.Yu., Ryabova, G.O. and Skripnichenko, P.V., 2015. The force model for asteroid (3200) Phaethon. *Planet. Space Sci.* 118, pp **296–301**.

[7] Skripnichenko P.V., Galushina T.Yu 2013 The study of perturbation structure and probability orbital evolution on the example of asteroid 99942 Apophis *Izv. Vuzov. Fiz.* 56 (6/3) pp. **229-231**.